War's Changed Landscape
A Primer on Conflict's Forms and Norms

И вот случился февраль 2022...
And then February 2022 happened...

PADDY WALKER
PETER ROBERTS

Howgate Publishing Limited

First published in 2023 by
Howgate Publishing Limited
Station House
50 North Street
Havant
Hampshire
PO9 1QU
Email: info@howgatepublishing.com
Web: www.howgatepublishing.com

British Library Cataloguing-in-Publication Data
A catalogue record for this book is available from the British Library

ISBN 978-1-912440-49-8 (paperback)
ISBN 978-1-912440-48-1 (hardback)

Paddy Walker and Peter Roberts have asserted their right under the Copyright, Designs and Patents Act, 1988, to be identified as the editors of this work.

The views expressed in this book are those of the individual authors and do not necessarily reflect official policy or position.

Contents

Dedication

Major General Patrick Brooking, CB, CMG, DL, 1937-2014
5th Royal Inniskilling Dragoon Guards
General Officer Commanding, Berlin, 1986

General Sir John Hackett, GCB, CBE, DSO & Bar, MC, 1910-1997
8th King's Royal Irish Hussars
Author, *The Third World War: The Untold Story*, 1982

Brigadier Harry Walker, MC & Bar, MBE, 1920-1969
5th Royal Inniskilling Dragoon Guards
Director of the Royal Armoured Corps, 1967

Thanks

To the 60-or-so contributors who provided this book's primary evidence base

... and

Lloyd Clark
Patrick Hinton
Kirstin Howgate
Madeline Koch
Olive Reekie
Angus Walker

... and

The Royal United Services Institute
The Humanities Research Institute, University of Buckingham

Preface

It is not accidental that a sub-heading for this book is '*And then February 2022 happened*'. That this is written into the title-piece in Russian is intended as a heavy-handed signpost to the effects on war of that country's brutal invasion of its neighbour. Indeed, several systemic factors, Russian and others, abruptly appear to be changing rules and norms of conflict previously thought immutable.

A portfolio of drivers seems to be accelerating this flux: Quick vacillation in Western public perceptions and the shaping of sensitivities by countries' media; a hastening in Great Power competition; a democratisation of weapons and their use; the blurring and hybridisation of participants' roles in battle; as well, of course, as the ongoing emergence of more and disruptive technologies. All of these factors give a new urgency to understanding the *degree* to which these influences are pushing actors to deviate from previously established behaviours in how conflict is undertaken. While a portfolio of new technologies may show promise in research laboratories (and, in so doing, offer militaries a pathway to a portfolio of advances across artificial intelligence and lethal autonomy, hypersonic weaponry, nano and bio engineering, and the like), several of these developments still require generational step change in capabilities before they can be deployed. Once introduced, however, they will posit a significant change in how humans wage war. The verso, of course, to this narrative is that Russia's invasion of Ukraine reminds us how *little* the battlefield really changes while, at the same time, the ethics, morality and legal frameworks for war's processes, old and new, lag far behind their adoption at both the military and political levels.

The authors' intention here has been to write a primer, a short introductory piece, the aim of which is to highlight that decisions on how to engage with these many developments are required well *before* policy choices are made over the remainder of this decade. While the authors use the artificial cut-off provided by the start of Ukraine's counter-offensive campaign in June 2023, the subject's enduring importance is that several well-tried concepts which have long comprised battlecraft may no longer be fit for purpose. Whilst commentators have long suggested that change is *the* constant in warfare, understanding the likely rules, norms and behaviours that might arise from these transformations is therefore generationally important. It is also doubly valuable during the period we find ourselves in where the language and nuance of the Western policy debate often seems frustratingly undeveloped. The primer's principal aim is therefore to raise the level of informed discussion in the topics that make up these debates and to do so right across policy-making domains such that those decisions can be rooted in evidence and the broadest possible range of experiences.

Prologue

February 2022 is an expedient, albeit horrific, stake in the ground that reminds us how long-held and popular narratives on contemporary warfare have actually been upended. Indeed, early evidence gathering for this publication, taken between 2019 and 2021 through interviews with sixty-or-so thought leaders from the military and government, from academia and from the third sector, now appears rather extraordinary in its almost unanimous agreement that future warfighting would be based upon all manner of means *except* a strong conventional fist.

This disconnect forms an important backdrop to the chapters that comprise this book and it is therefore useful to signpost the reader on the topics that make up the authors' analysis. As will so often be the case in this primer, context is key to understanding the pace and degree of change. In considering, for instance, how Russia might fight its war in the Caucasus or, indeed, anywhere else, Western military experts used context as the anchor to predict an impressive demonstration of military might in which a modernised Russian military machine would operate seamlessly in the multiple domains of air, land, sea and cyber. And while it would undertake this using regular military assets, war's new character would primarily be defined by clever, unexpected combinations of irregular paramilitary forces employing a broad, bewildering collection of asymmetrical means.

This has not played out. Instead, President Putin's expectations for a largely uncontested conquest have been undone in its early phases by absent planning, bad logistics, poor communication security, and the amateur use of armour in complex and foreign terrain, and by tasks being undertaken with virtually no air-land cooperation. It has been hamstrung by insufficient infantry who are anyway poorly led, as evidenced by widespread counts of ill-discipline. Russia's initial playbook in Ukraine has been shown up as a bad plan with no defined effort. But, for the purposes of this primer, it has also revealed a set of assumptions and premises about how belligerents will fight, and about what behaviours, forms and norms they will apply on the battlefield.

The primer therefore seeks to outline the factors driving these behaviours and, as such, it is less about a seemingly first-world military power humbled by a much weaker adversary (history is full of such examples) and rather about the manner and means involved in that chastening. It is informed by the wide array of failures, the forecast outcomes that have been undone and where hitherto adopted models have been found wanting. Finally, Russia's experiences in south and east Ukraine have also refocused eyes upon conflict's soft factors, upon newfound doubts around technology and, fundamentally, what comprise the *empirics* of winning on the battlefield.

Trying to derive lasting lessons and consequences from recent campaigns is always fraught with challenge. Systemic changes in conflict's character were, after all, already plain before the Kremlin's decision to cross into its neighbour. Technical disruption, broad developments across fighting domains and a decade of profound societal changes had already complicated how wars were to be fought. But this all has a series of versos (a construct that is repeated throughout this primer), and, after all, the Ukrainian conflict is but one war. It involves just two adversaries and, in the case of Russia, commentators have long found it challenging to pinpoint didactic behaviours and actions, itself an important component of that country's toolbox to surprise. Nevertheless, it would seem reasonable to suggest that the changes wrought by Russia's wider actions must have material consequences on behaviours and conventions. Previous shocks, after all, have unwound by ushering in new eras and several metrics would suggest that Russia's actions will occasion similar. As at June 2023, nearly 12 million people had been displaced, the majority of those leaving their country pushing past the first border they encounter before moving to adjacent or more distant states. All told, nearly 60 per cent of the six million-or-so actual refugees are now housed somewhere beyond Ukraine's neighbouring countries.[1]

---

[1]   McKinsey & Company, Occasional Paper, 'War in Ukraine: Twelve Disruptions Changing the World – An Update', 29 August 2023.

Other disruptions must inform ongoing norms. In the energy space, for instance, Europe has doubled its imports of non-Russian natural gas in the 16 months since the invasion while reducing its consumption of the raw material by more than one tenth. Defence spending is rising, especially in former Eastern bloc countries. The war has generally occasioned seismic changes in practices, from corporate actors pulling back from doing business in Russia (of 283 non-Chinese Fortune 500-or-equivalent companies transacting in Russia before February 2022, just 17 continue overtly to trade in that geography as at June 2023) to cyber and conventional forces combining for joint attack. These are all reasonable change agents that should be factored into assessing changes to conventions.

A key purpose for this book has therefore been to review whether norm change is taking place at a rate hitherto unseen and to undertake this analysis in light of recent advances in technology, of those same shifts in societal attitude, as well as with regard to degrees of disruption in actors' current strategic and risk calculi. Importantly, this purpose remains unchanged notwithstanding current events in Ukraine. It still concerns warfighting and the meld of means that this entails. It involves context and bringing balance to wild claims around new practices and suggested discontinuities. It is about technology, leadership, pace and endurance. How are operations and the forces that undertake battlecraft to be managed at scale when warfare nowadays can be conducted at machine speed? Indeed, how broad and how relevant is this catalogue of factors when examining near-term norm change? After all, analysis making bold, far-reaching forecasts may make bigger waves, regardless of how correct those forecasts later prove, but while this book may suggest a new speed characterising the pace of change in war's character, it is incrementalism that remains the overarching norm in war's prosecution.

Other matters complicate this analysis. President Putin's initial framing of his conflict as merely a 'special operation' by very definition limited the tools and tactics that could initially be used from his arsenal and, as 2023 draws to a close, it still remains unclear how the hand of cards deployed by Russia's leadership will play out. For the purposes of this book, moreover, *political* considerations in Russia so dwarf its military playbook that predicting the conflict's enduring effect on warfare's norms becomes twice as unclear. Setting those challenges aside for a moment, whether in warfare's means or in the vicissitudes of that conflict's geopolitics, the relevance of current norms becomes very apparent, the more so given the deep surprise occasioned by Russia's brazen reliance on *conventional* methods.

'Surprise' would seem to be an odd characteristic to highlight in today's era of ubiquitous surveillance, intelligence and seeming knowledge. Two observations arise. First, it should shock no one that the most impressive policy prognosticator gets many things plain wrong. Our world, both socially and politically, is enormously complicated. Second, models and theories are extraordinarily sensitive to their underlying assumptions and these, of course, are more often posited than proven. Nor, it turns out, does an emerging age of big data provide clarity to these age-old challenges. Indeed, it is the diaphanous nature of norms throughout this analysis that makes meaningful attribution particularly challenging. How information is now disseminated and consumed turns out to be a pivotal and accelerating driver in shaping norms to the degree that the authors give the matter its own chapter in this primer. Indeed, Ukraine demonstrates the gulf between merely garnering this information and deriving relevant understanding from those processes. It turns out that scraping, handling, manoeuvring and moving around data at breakneck speed do not equate to insight. Nor do they equate to divining awareness or meaning, the more so (it turns out) when procedures have been delegated to machines without meaningful human oversight.

Given that purported 'information advantage' has long been a battlefield staple, it may appear that the authors spend disproportionate time in later chapters on tracing, for instance, the impact of the smartphone, its tools and ecosystems as well as the recent enabling effect of connectivity. It is, however, exactly this 'new' information that so materially shapes parties' actions, informing the complex equations that comprise norms of warfare. But this is also a complicated dynamic, especially given that behaviours are clearly perceived quite differently between the West and others. Insight and sensitivity rarely travel well across borders. A second task therefore becomes one of judging the pace and ramifications of this divergence, best captured by Elbert Hubbard's observation made more than a century ago that 'the world is moving so fast that the person who says it can't be done is generally interrupted by someone doing it'.

Introduction

This book is structured as a primer. It seeks to establish a short, readable and non-technical baseline on likely norms and forms of warfare over the coming 15 years and to do this from a single point in time. It also undertakes this exercise from a position of first principles, focusing on concepts and notions in order to help the reader with an understanding of the many components that underpin norms and their movement. The book generally tries to tease out higher level theories and their assumptions rather than build its argument upon specific examples of, say, the latest weapon systems to hit the battlefield or other new forms of warfare which may or may not be relevant in the years ahead.

The book's strapline is 'A Primer on Conflict's Norms and Forms'. Here, *norms* of warfare concern patterns of behaviour which make up the rules and often ambiguous responses that drive actors' actions and responses. The *forms* of warfare then concern the measures and means undertaken by both state and non-state actors in the activity of *prosecuting* war. War's norms and forms are necessarily intertwined. Norms govern the behaviours and conduct of actors. They are generally the extension of existing legal frameworks and other rules-based arrangements that underpin today's international systems. While certainly subject to violation and local interpretation, they have traditionally been stable and enduring and are important precisely because they act as a means of ballast to balance-of-power arrangements and the previous resort to brute force and coercion in order to solve state-level problems.

The Norms of Warfare

The primer packets norms into three separate, albeit overlapping categories. First, *enduring* norms relate to existing, long-dated and persistent behaviours. They represent the immutable practices that underpin the conduct of warfare. They are largely transactional, meaning that they attach to activities that comprise adversaries' battlecraft, the processes that together constitute the waging of war. Their longevity and durableness mean that the primer often uses the adjective 'current' when discussing their contribution and place in today's wider debate. This should not imply, however, that any battlefield behaviour can be immune from change over the period of this primer's consideration. All aspects of war's conduct, after all, are constantly placed in the crosshairs of change by developments in war's forms (the assets and wherewithal at commanders' disposal to wage war and the *means* of warfare that then arise), suggesting continuous adjustment over time as these norms shift to account, for instance, for emerging technologies or other new battlefield circumstances. Enduring norms, however, generally remain absolute, either as customs that are isolated from change by being tied to war's unchanging nature or, in the case of changes to war's character, sufficiently broad for those downstream changes to have neither material nor immediate effect on their scope. They are abiding and systemically stable in their nature and unlikely to change over the 15 years under consideration by this analysis. Indeed, a conclusion of these chapters is that these conventions are generally *slower* to evolve than is commonly thought. The grounds for this are plentiful and include, inter alia, general frictions from inertia and integration processes, the perennial phenomenon of plausible deniability, the continued challenges of logistics and resupply, matters of persistent competition and operational surprise, and, just one component of a long list, behaviours arising out of legitimacy and motivation.

The second category of norms considered here relates to those that are *emerging* or *evolving*. These still concern patterns of expected conduct and conformance but, in this guise, relate to a somewhat distinct set of developing and progressing behaviours that have been occasioned by recent advances. They remain in flux (and hence the use of the qualifier 'somewhat' in describing their current state of morphing at any point in time). Examples include changes in behaviour arising from the erosion of stability mechanisms, and from the rare deployment of properly disruptive new means of fighting, from the ever-widening definition of war that will be discussed in later chapters, as well as from the control of narratives and the dynamics of peer, near-peer and previously non-peer relationships. Generally empirical, they relate to still changing characteristics of warfare and, today, are usually both occasioned and framed by emerging technology and the new capabilities that it creates. An example here might arise from the incremental re-engineering of command processes to match the introduction, for instance, of remote weapons and other technical innovations in delivering lethality on the battlefield. While their implications may certainly be significant (both around whether current systems remain sufficiently robust to meet near-term challenges but also around the West's moral, ethical and legal frameworks and, in this case, their continued relevance as available means of warfare multiply), they generally have less precedent upon which to anchor analysis. For the purposes of this primer in 2023, it remains unlikely that they are to be sufficiently coalesced to be defined either as a concrete or wholly new norm of warfare. For the purposes of this book, *evolving* norms therefore represent a likely long-standing pattern of behaviour that is still undergoing material change but with its basis informed by earlier understood actions that serve to anchor both its foundation and broad definition.

Finally, a *new* norm may be defined as a behaviour or rule set that has undergone significant recent change but is now cemented and understood going forward as having captured a permanent alteration in war's conduct. Examples here might include behaviours arising from a particular expansion in the means and forms of warfare, the discontinuity occasioned by ubiquitous connectivity and the consequent rise of the 'digital individual', as well as material impacts upon battlespace from data and its ramifications. By definition, new norms are young and less stable. They owe their creation to sets of circumstances that themselves may then change. A further point is worth making. The complication here is that their advent has reflected and captured a particular new situation, factor or set of events that must then in turn be echoed across war's other norms. The arrival of a new norm might be thought to bring new constancy to battlecraft. It rarely does.

These three distinctions may also appear only mildly helpful as norms' foundations defy easy abstraction, and are dynamic and liable at the edges to quick and unexpected change. However, principles of right action have customarily been underpinned by pragmatism and it is traditionally actors' basic rationality that encourages norm adherence as an attractive stabilising influence on their behaviours. To be aggressive, after all, is to violate a norm of war notwithstanding parties' efforts to frame aggression as a legitimate means of defence. This primer must therefore consider a very broad range of possible change agents to passing norms regardless of type and to do this in light of a portfolio of long-dated, systemic challenges that otherwise tend to stymie developments in behaviour: how to reintroduce mass onto the battlefield; how, perhaps, to reduce the cost of power projection; how to combat specific new technologies; how to deal with the issues of resilience, replenishment and buffers; and how to update doctrine to make it fit for purpose in this era of certain change. A list here is useful if only to evidence the catalogue of issues that must feed into norms' consideration and dynamic definition.

The Forms of Warfare

The 'form of war' is an altogether easier notion to wrestle with as it is fundamentally just another way of addressing *how* (the means employed) a force will fight an adversary. Forms concern *how* parties fight ('go hard, go fast, go home'). They are not about doctrinal underpinning and instead concern operational art, tactical science, wherewithal, allocation resource, development of new battlefield means and, generally, the deployment of available assets in theatre in order to achieve intended outcomes. Forms of war are the tangible product of strategy, the means of undertaking the requirements thrown up by strategic process and, therefore, executing the end-product of Grand, Political and then Military strategy.[1]

Technology may be a part of this but it is also about operational design, assets and, generally, the *character* of how war is undertaken. It concerns the plans, designs and means employed, the rules and boundaries around how it is envisaged that force will be deployed and, ultimately, the amount of 'blood and treasure' that a belligerent is willing to invest in that war. War's forms are essentially mutable, ever developing and adjusting to seek advantage. They also concern the phasing and degree of conflict, the availability of people and materiel with which to undertake the fight and, of course, the aims that one is fighting towards. One definitional difficulty arises. While forms may appear pigeonholed by the robust, identifiable factors that comprise war's means, the degree to which these forms are significant (and, more importantly, the intensity with which they go on either to alter or create war's norms) still depends upon nuances arising

---

[1] Francis Miyata, The Grand Strategy of Carl von Clausewitz, 26 March 2021, *War Room*, US Army War College, https://warroom.armywarcollege.edu/articles/grand-strategy-clausewitz/.

from how these assets are used and the effects that they achieve. These similarly defy abstraction and depend, for instance, upon forms' intended aim either to win the fight through the complete destruction of an enemy force or just its coercion to the point of their defeat.

War's forms are therefore more akin to its characteristics. Just as norms provide the broad framework through which war's forms may be understood, those forms relate to the methods and practices that define the passage of conflict. They are empirical, observed and changeable. Every aspect of war's undertaking has its own such form, from war's tactics to commanders' allocation of available assets. It is, moreover, the *amalgam* of forms that then defines war's character. War's forms are then endlessly evolving, whether through political direction, through shifting priorities assigned to them by military command or, more prosaically, depending upon local availability of materiel, technology and other enabling means. Forms are the visible manifestations of warfare, the series of happenings and their drivers each day, each hour as battle unfolds.

Indeed, once deployed, these forms then occasion developments in either evolving or new norms which individually may then become recognisable components of war's passing character. Examples abound including the 'shock-and-awe' of the first Iraq war. A better case is provided by Russia and China's use of a portfolio of effects across political, military, constabulary, economic and legal domains which, once aggregated, repurposed and honed into new practices (to include, for instance, the sub-threshold, electronic and societal) then morphs into its own defined new means to undertake warfare.

Methodology and Assumptions

This book's primary research source comprises more than 60 interviews and written submissions drawn broadly from the UK and US Forces, from academia and those in politics, and from those in think tanks and the third sector.[2] These were typically hour-long, semi-structured interviews undertaken on a non-attributable basis but generally informed by a series of questions intended to make sure that the relevant research question was properly aired and its sub-topics covered. The talks took place in

---

[2] Third sector organisations here refer to charities, non-profit organisations, think-tanks and other non-governmental bodies with special interest and skills in their relevant vertical. An example is provided by NGO Human Rights Watch and its actions around munitions use, human rights monitoring and, generally, oversight of activities in combat zones.

2020 and early 2021, the content deliberately focusing on norms' concepts and constituents (rather than specific passing examples that again may or may not be relevant in decades' time). Those participants are listed as an appendix to this book. Given subsequent events, the material captured is doubly interesting both for what turned out to be its poor calibration in relation to unfolding events in Ukraine and also as a record of deep-seated passing orthodoxy in the months before Russia's invasion of its neighbour. A synopsis of this primary evidence is included in a further appendix as an aide-memoire to encourage supplementary debate as well as to provide additional colour to the arguments made in the book by its authors. The summaries are also intended as a teaching aid for those interested in developing further the arguments made by the authors. The book, after all, is deliberately a snapshot in time and a further point to the summaries is therefore to provide a baseline and reference set that will help in the ongoing updating of the book's analysis.

The book has also been informed by contributors to RUSI's podcast, *The Western Way of Warfare*, and other adjacent sources. While much of this evidence was collected *before* the geo-political events of February 2022, it was still compiled against the backdrop of considerable transformation in how UK forces may fight.[3] This is well evidenced by the inaugural RUSI lecture of the then incoming Chief of the Defence Staff in December 2021, three months before Russia's invasion of Ukraine: 'It is now clear that the last 20 years were not the end of history. At best, it was a pause… The State is back with a vengeance. Indeed, for our competitors, it never went away.'[4] This is a foundational observation that informs much of this primer's conclusions as a snapshot document that is intended to act as a catalyst for further debate into what might further shape those norms as the UK enters the second and third decades of the 21st-century.[5]

---

[3] See UK Cabinet Office, 'Global Britain in a Competitive Age: The Integrated Review of Security, Defence, Development and Foreign Policy', 16 March 2021, https://www.gov.uk/government/publications/global-britain-in-a-competitive-age-the-integrated-review-of-security-defence-development-and-foreign-policy.

[4] Admiral Sir Tony Radakin, *CDS Annual Speech*, Royal United Services Institute, 7 December 2021, https://www.gov.uk/government/speeches/chief-of-the-defence-staff-speech-to-the-royal-united-services-institute, and *Armed Forces to Be More Active Around the World to Combat Threats of the Future*, 23 March 2021, Gov.UK, https://www.gov.uk/government/news/armed-forces-to-be-more-active-around-the-world-to-combat-threats-of-the-future.

[5] Generally, third party journals and online material are credited. Commentary from the project's initial commentators is typically not credited but informs the primer's direction and conclusions. Online resources were accessed between March 2022 and June 2022. In addition there is a synopsis of the project's primary evidence, included without attribution.

Again, the reader should note that the book is informed more by concepts rather than by particular happenings or battlefield examples which, over the three years' writing of this book, seem individually to be less relevant and quickly out of date. Indeed, the working evidence provided by the book's early interviewees may appear inconsistent, slightly erratic and sometimes contradictory, but is valuable precisely because it is derived from ideas, notions and conflict's high-level design rather by a series of specifics driven, perhaps, by a new weapon variant, a new means of engagement, or the like.

This also holds true in the primer's handling of war's purported phases. Theorists' views on war do not necessarily chime with realities on the battlefield. Regardless of the terms that are applied to war's practices, adversaries are just trying to do their best and actors are generally motivated solely to do what is in their interests at the passing moment. This book, therefore, may not be the best place for insights on positional or manoeuvrist or attritional plays. It does touch on attack and defence, but it does this from a position that warfare is generally situational with little concrete difference between these and other states of the battle. Attacking your enemy's cohesion and their will, avoiding their strengths and trying to leverage their weaknesses are age old means of warfare and tying them to new terms and labels does not constitute their new means to win the fight. Context and situation are usually the ingredients that decide the degree to which these three principles are ascendant and whether that day's battle is positional, roving, manoeuvrist or being undertaken through attrition.

All elements of warfare are anchored to some degree by these principles and, while each of these states certainly constitute phases of the battle, it is context that dictates which particular principle is relevant at any particular time. The issue is therefore approaching battle with a playbook and labels that are over defined and insufficiently inflexible. A long-dated norm is that types of fighting are themselves fungible. They change rapidly, and are mutually interchangeable and replaceable. Engaging in urban warfare and generally in the close fight provide useful examples. Commanders may have the *components* to wage the war that they would like only to be frustrated instead by their immediate conditions (their proximity to the enemy, the absence of opportunity, changing priorities that are forced upon them by what is happening in their near environment). Enemies, after all, will not fight in the way that this commander (and their dogma) expects and it is what is immediately happening on the ground that prevents actioning any full set of options that are laid out in one or other of these doctrines.

Phases of battle are difficult to abstract. In this vein, the norm should really be that attritional warfare is actually less a discrete and deliberate series of actions and is instead more a characterisation of the battlefield and its environment at any moment of the fight. The observation is mutable across battlecraft. Context here is all about the fighting's environment and the degree of contact with the enemy, and not about elegantly defined modes of engaging the enemy. The weaker party, for instance, will move back into an urban setting to offset the strengths of the stronger adversary; this is a positional move but also, confusingly, an example of the opportunistic approach of the manoeuvrists.[6]

Finally, the analysis is then bookended by an artificial cut-off provided by the start of Ukraine's campaign in June 2023. For the reader, therefore, the book's observations are informed by the general experience base of those interviewed, adjusted for outcomes and lessons taken from the first 14 months of Russia's campaign in order for the authors to comment on how battle might be undertaken over the next decade and a half to 2040. Or thereabouts.

The book's research methodology has been to crosscut the primer's main title into subordinate themes, each of which provides stand-alone chapters but which should be considered in the round in order to achieve the intended baseline of understanding on the topic. In so doing, the authors first consider the role of context and prior art in assessing likely changes to norms over the period. It is not, after all, technology alone that can fundamentally alter the nature of the future battlefield. To achieve proper traction, any advance, technological or otherwise, unsurprisingly relies upon human endeavour and leadership, upon enabling infrastructure and the delivery of other (possibly still undeveloped) capabilities if it is first to be integrated and then have the systemic impact that is required to affect norms. Indeed, the success of novel systems in legacy force design is directly proportional to ideas and human execution of those ideas. Moreover, their depth of adoption is contingent upon navigating both legacy enablers but also their barriers, all of which shape passing battlecraft.[7]

While this relationship may present opportunities, it also creates unseen vulnerabilities requiring management and prioritising if new forms are to be deployed and, in time, norms perhaps impacted. This is

[6]   Amos Fox, 'Is Manoeuvre A Myth?', *This Means War* Podcast, 24 August 2023.
[7]   Gabriella Blum, 'The Paradox of Power: Changing Norms of the Modern Battlefield', *Houston Law Review*, Vol. 56, No. 4, 23 2019, https://houstonlawreview.org/article/7948-the-paradox-of-power-the-changing-norms-of-the-modern-battlefield.

not straightforward and debate around the relative merits of particular battlefield assets have long been fierce among those observing how battle is being fought and the progress one or more adversaries may be making at any point in that conflict. Much of such attention, however, takes place without appropriate grounding in data. In particular, statements on capabilities are framed either by the unlikely claims of manufacturers, on procurement's wishful thinking or, more usually, in response to the information manoeuvre of other interested parties. In considering novel systems, therefore, whether in weaponry or in how organisations develop, the primer attempts to reflect instead on how norms may be shaped by changes in how operations are undertaken, the effect and challenge, for instance, of loosening human supervision across lethal engagements, the fast-evolving role of the battlefield's electro-magnetic space and how leaders must adapt to these new circumstances. The primer then concludes by considering the place and function of the laws of armed combat and human rights, as well as leadership ramifications of ethics in future conflict.

The lack of dependable information on parties' dispositions, decision-making, and their wins or setbacks makes deducing lessons from recent war zones an ever-riskier task. This also has consequences for the primer. Sketchy data, the likely selection biases within that data as well as the testimony-based and unattributable nature of the book's primary evidence, mean that the primer is not intended as a work of academic scholarship. Instead, it should be taken as a compendium intended first to identify, then to challenge and, finally, to substantiate behaviours and their movement as they relate to the battlefield over the coming two decades. Little, for instance, is reliably available on what really comprises adversaries' capabilities and the behaviours and intentions that might be expected from them. Despite commentators' apparent confidence, the degree and impact of those combatants' *forms* of warfare unsurprisingly remain an unknown. This is especially the case around warring parties' more ambiguous and sinuous activities which, by definition, are undertaken in the shadows and usually under a cloak of plausible deniability as opposed, perhaps, to their conventional assets which appear more observable for analysts to dissect. Finally to this point, constant developments in war's forms as well as both the degree of surprise and speed of these changes make it more useful for the authors to publish an imperfect and likely transitional study rather than wait for more authoritative lessons to be drawn (for instance, concrete lessons from the Russia-Ukraine conflict that goes on as this book is published).

Given its reductive nature, therefore, the primer must make several assumptions. First, the 'Western Way of War' has increasingly become the 'American Way of War'. Notwithstanding events in Ukraine, this still has had two adjunct effects on current norms. First, other actors, over recent decades, have been playing catch up. But second, there has been pressure on parties to trial new methods and, again by extension, establish new norms in their efforts to achieve (or exceed) battlefield parity with their likely adversaries. Both developments represent trends that have also been led by ideas (and not, counter-intuitively, just by technology) and in lockstep with a growing realisation that there are many ways to win at war. This empirically continues to encourage variation in actors' activities and, in such manner, to a portfolio of possible norm-changing behaviours. It is widely noted, after all, that many more components of statecraft and states' activities can now be weaponised, a recurrent theme to later chapters of this book.[8] Here, a further key driver for this primer is that a large number of small adaptions can quickly add up to fundamental change. While this tenet might traditionally have quickened in times of war (in direct response, for instance, to the actions of adversaries), the evidence taken for this primer points repeatedly to a new norm of *generally* faster change and quicker timelines.

All of this is also prompted by that return to great power competition identified in the Chief of Defence Staff's December 2021 speech. What might this mean in practice? In normative terms, it underlines again the importance of readiness and posture, a re-emphasis upon national will, adjustment to previously presumed fighting concepts and innovative operations that have been properly integrated into legacy practices. It underscores the values of fluidity and flexibility, whether that be in the rapid adaption of off-the-shelf technologies or delays to the retirement of older platforms. Indeed, a long-dated and enduring norm remains that new practicalities always arise that must be re-mastered and tuned by each competing adversary. Ukraine's defenders and their willingness to embrace new means in face of such grim attrition provide a case in point. But while each of these initiatives can appear transformational in itself, the primer's conclusion will be that individual developments rarely seed a new norm. This requires time, first to identify and then process the aggregation of effects arising from individual developments. It also requires the factoring of intangibles; as demonstrated

---

[8]   Henry Farrell and Abraham Newman, 'Weaponised Interdependence; How Global Economic Networks Shape State Coercion', *International Security*, Vol. 44, No. 1, Summer 2019, Executive Summary and 42-79.

in Ukraine, it is the *will* to fight that remains the key component in deciding whether an adversary will triumph in any phase of battle.

A contentious issue for this analysis is therefore how the West, militarily, academically and politically, read this set of events so poorly, in particular its collective failure to recognise the ongoing place of conventional warfare as a means for politicians to achieve their ends. Much of this is to do with the trends that currently shape debate in Western military thought: innovation has become preferable to evolution; data analytics undertaken by machines has overtaken that provided by human analysis; the *human* cost of conflict has become increasingly unacceptable in Western states. Together, these factors have driven Western doctrine towards an *idealised* form of warfare without bloodshed, the dichotomy that exists between parties' embrace of new means but in a context of limited investment and short termism. It should not therefore be surprising that the promise of technology as an attractive means of solving conflict's problems (with its potential for precise, clean and remote fighting) has become a central tenet of Western military thought. This has then been accelerated by a further happening in Western capitals, with the widely held hubris to believe that adversaries might then fight according to these Western rules. Indeed, norms lend themselves to being considered along the line of a continuum, at one end of which is the uncomfortable contention that norms are a luxury, the preserve of Western democracies to be ignored (or certainly subverted) by autocracies or those parties who will seek advantage regardless of expected rules and behaviours. This is considered in later chapters.

A purpose of this primer then becomes the creation of a ready resource that encourages future force designers to avoid preconceptions, groupthink and the thin slicing of history. It is intended to highlight the dangers of presentism, neophilia and the failure of challenge. It is about reviewing norms in light of current context, navigating better a direction of travel in warfare's means and, in so doing, to understand better why Ukraine's 2022 invasion is such a wake-up call for the West. In order to achieve this, the book's central inquiry becomes whether forms of war and warfare will be predictable in the future, so aiding future force designers' shaping of militaries devised to protect and defend in this future.

Context in Norm Analysis

Understanding norms and their constancy requires consideration of context. The exercise relies upon deciphering between warfare's trends, perhaps quite short-dated and abrupt, and the moment of cross-over from trend into permanent change which then might herald, through adjustment and development, an evolving or even new norm of warfare. The analysis is complicated precisely because the new norm is fluid and defy abstraction. Norms lack neatly defined edges. A recently stated priority for UK forces, for instance, is that they be both 'more deployable' and 'deployed more'.[9] This might suggest a new modus operandi and, in time, perhaps an evolving norm (whereby, presumably, UK assets are used in support of UK interests, deterring UK adversaries and attempting to shape matters on a continuous basis). Using context, however, as the lens through which to judge the degree of systemic change provides the key filter and suggests, of course, little concrete change of actions aside from a reshuffling of duties and further trade-off. Instead, it is driven by politics and headlines and the wish of policymakers to dress incremental change as something altogether more seismic.

A recurrent theme therefore quickly emerges that little really changes. Indeed, to the extent that events do suggest change to a norm, that norm empirically tends to become more and not less stringent. Similarly, occasions of norm violation tend to seed self-correction in that norm. Norms can carry along unchallenged for an age until some contrary occurrence prompts an abrupt re-evaluation of that behaviour, usually cementing the foundations and assumptions upon which it is based. Categorising, therefore, where a trend has become a new norm requires patience and speaks to the snapshot nature of this primer. Trends, fundamentally, may mutate quite quickly and, as such, are not necessarily harbingers of norm change. Today, for instance, conflict's fleeting characteristics might include the role of irregular warfare in deterrence, the shape and means of integrated deterrence in an era of strategic competition and, at the granular level of the battlefield, the tying-in of information operations with kinetic capabilities. All of these are certainly important developments but perhaps individually belong further along this same continuum and where readers should not expect resolution for some time. Tomorrow's list of candidate norms might be very different, and it is this general lack of definition for norms and their framework that adds challenge to their forecasting.

---

[9]   Admiral Sir Tony Radakin, *CDS Annual Speech,* Royal United Services Institute, 7 December 2021, https://www.gov.uk/government/speeches/chief-of-the-defence-staff-speech-to-the-royal-united-services-institute, and *Armed Forces to Be More Active Around the World to Combat Threats of the Future,* 23 March 2021, Gov.UK, https://www.gov.uk/government/news/armed-forces-to-be-more-active-around-the-world-to-combat-threats-of-the-future.

This primer must also look to establish the *degree* of norm transformation given the matter's fast-moving strategic context. It seeks to do this through a global prism but, as appropriate, with particular regard to the United Kingdom. To what degree has the set of circumstances facing the broad UK defence establishment changed over the past half-decade such that transformative trends are 'bringing the future faster'? As importantly, how has Russia's invasion of Ukraine altered the set of previously accepted equations that drives UK posture? Just as it is difficult to pigeonhole changes in the ways that humans go about their business, the point here is that it is the portfolio and *cumulative* nature of emerging trends that are relevant to how current norms and their framing must be considered. As the primer is a snapshot in time, this task is also complicated by difficulties in prediction and, in an age of democratised technical upheaval, the contextualising of second-order effects impacting what were previously considered foundational, immutable practices. A case in point is the upheaval in social media and its refashioning of social norms. Today's social media tools, after all, will look less and less familiar as the book's 2025-2040 timeframe unfolds. Similarly, the characteristics of war that are observable in Ukraine will require that planners consider future norms against a portfolio of very different scenarios, quite distinct from the playbook being followed by Russia's invading forces. Twain's maxim about history not repeating but rhyming is a useful yardstick for this primer.

Changing Norms around War's Character

Second-order effects complicate attribution. Nowhere is this more evident than in the world's adoption of social media, where ubiquitous and easy-to-use tools now allow everyone with a phone to disseminate views. The whole population suddenly has the wherewithal to become an activist with previously unforeseen reach. In considering norms, this is disruptive precisely in how manipulation of information is newly facilitated for a very wide cohort of parties. It is social media that enables significant interaction (and, by extension, 'interference') in matters which previously appeared anchored and less fractured. In assessing permanence, it is also clear that these tools cannot now be uninvented. Notwithstanding the many versos to this phenomenon (output, after all, from these tools does not revolutionise parties' ability to understand what is happening, as evidenced by Ukraine's ability to maintain effective news blackouts during sensitive operations), these tools will continue to morph, complicating defences on how they may influence behaviours. In the states' sphere, this has systemic ramifications but ones that are difficult to pinpoint, whether through social media's far-reaching role in a new-found touting of nationalism ('Buy American'), a revived emphasis generally on near-shoring or a return to foundational rivalry, both for assets and for ideas. In the public realm, it may be the sowing of unrest or the manipulation of narratives, both tactics informing a very strong (although complicating) social context, social behaviours and, by eventual extension, pressure upon passing norms.

On both sides of the conflict, Russia's invasion of Ukraine evidences the importance of information as a weapon of war. As with many of the observations in this analysis, there is often a verso to the conclusions that the authors draw out, this time regarding state-controlled narratives. Social media has wholly compromised governments' ability to mould narratives and prevent deviation from official storylines. A second verso arises. Open-source intelligence has quickly become a generally workable tool for parties to corroborate information and fact-check narratives. It has also developed into its own means of warfare whereby such newly ubiquitous data, in this case from the internet, from satellites, from web cams or from civilians' mobile phones, can now be collected, processed and manipulated with less and less friction. As noted by the *Economist*, 'a photograph of any spot on earth, of a stricken tanker or the routes taken by joggers is available with a few clicks'.[10] Militaries, experts and hobbyists alike can now create, check or repurpose information, whether to derive intelligence, solve riddles or reveal misdeeds with unprecedented speed. The tools turn poachers into gamekeepers but also gamekeepers into poachers, levelling a previously uneven playing field and so upending passing behaviours.

Indeed, this all has enduring ramifications for how war will be undertaken. In Ukraine, for instance, this has enabled significant new targeting practices on the back of new data access, collation and then analytics, all signposting important emerging norms, the more so given the phenomenon's far-reaching consequences. Information, after all, can also promote transparency, strengthen law enforcement and, in facilitating that corroboration of versions and interpretations, is likely to have an enduring (although not yet defined) effect on war's very conduct. Here, therefore, it is the *democratising* of information that is the likely discontinuity. It is also one

---

[10] Economist editorial, 'The Promise of Open-Source Intelligence', *Economist*, 7 August 2021, https://www.economist.com/leaders/2021/08/07/the-promise-of-open-source-intelligence.

that serves to blur previously distinct edges. Similarly, it is its decentralised and egalitarian nature that may also erode traditional power bases by laying bare 'previous arbiters of truth and falsehood, in particular', notes the *Economist*, 'governments and their spies and soldiers'.[11]

Again, however, it is difficult to be certain how norms have jumped in direct relation to developments in Ukraine over the 14 months to June 2023. New practices, after all, rarely enjoy more than fleeting advantage on the battlefield. They represent adjustments to the means of undertaking the fight, war's changing character. Practices and means are also very intermediated; they come together only through the participation of very many and usually quite disparate parties, partners and alliances, and their effect upon war's character is correspondingly muted. They are rarely disruptive. In the case of private military companies, the impact of Russia's Wagner organisation demonstrates that these cohorts still require state logistics, and still need its support and acquiescence in order to thrive.

Pitfalls to Norm Attribution

Pitfalls, moreover, exist in considering revolutions in methods and, in particular, the gauging of cause-and-effect from upcoming technologies and new processes. Here, Gray makes useful observations for those undertaking such analysis, many of which are foundational for this primer.[12] First, war's political, social and cultural context cannot be ignored. Second, defence parties prepare for problems that they prefer to solve rather than those a cunning enemy might pose. Trend spotting, after all, is fundamentally tricky; trends come in bunches and, together, it is their cumulative consequences that meld norms and shape future behaviours. Third, understanding this blend is an art and not a science and, finally to this point, surprises happen and must be calibrated for before conclusions can reliably be drawn. Military analysis is not alone in often being driven by 'presentism', the universal human condition whereby commentators consider their time as a period of unprecedented turbulence while simultaneously attributing an exaggerated tranquillity to the past.[13]

---

[11] Ibid.
[12] Strategic Studies Institute, '*Defining War for the 21st Century*', SSI annual strategy conference report, 2010.
[13] Jonathan Wichmann, 'Our world is changing – but not as rapidly as people think', *World Economic Forum*, 2 August 2018, https://www.weforum.org/agenda/2018/08/change-is-not-accelerating-and-why-boring-companies-will-win/.

The factor's importance is captured in the suggestion by Barnes that presentism and neophilia (here, the belief that what is observed and experienced in the battle space is entirely novel) tie militaries 'into an interminable and esoteric debate about the nature and character of war that is informed by budgets and reputation more than by history and experience'.[14] Indeed, this is too often borne out by the language used to describe developments in the space. Concepts such as 'new' and 'hybrid' war, 'un-war' and 'non-war' suggest that there is an important, possibly profound, distinction between wars of the past and those of the present and future. The description of war as 'hybrid' is a particularly overused term proposing near-term replacement of conventional force-upon-force by 'alternative' means of waging war that revolves around so broad a portfolio of means as to make the term meaningless; here, hybrid can cover activities involving cyberattack, terror and misinformation campaigns that happen without attribution (the notion of 'plausible deniability'), and the engineering of economic and other governance crises from other ambiguous and weakening activities.[15] Right or wrong, it is this wide mix that has provided a traditional underpinning of analysis into war's causes, means and assumed norms. A tighter definition for some of these ingredients, a subject for later chapters, will therefore provide better footing for this analysis as well.

There are two ends of this argument, the framing of which also bookends subsequent chapters of this primer. A norm, after all, is that a party (be that the traditional notion of a state or some other relevant agglomeration capable of waging war) may be at conflict without knowing it. Indeed, longstanding behaviour is that the start and finish points in conflicts have become increasingly imprecise with changes to the boundaries of what constitutes conflict, making it increasingly difficult to differentiate its military and non-military dimensions. Traditionalists, by contrast, occupy a space further along on this continuum and might disagree with such a hard and fast conclusion. They focus instead on continuity and consider conflict with a longer lens. Strategy, they contend, should not succumb to

---

[14] Paul Barnes, 'Neophilia, presentism and the deleterious consequences for Western military strategy', *Modern War Institute*, 3 June 2019, https://mwi.usma.edu/neophilia-presentism-deleterious-consequences-western-military-strategy/.
[15] Tracy German, 'How Will Wars Be Fought in the Future?', *Oxford University Press Blog*, 20 July 2019, https://blog.oup.com/2019/07/how-will-wars-be-fought-in-the-future/.

fads.[16] Instead, the relevant context is that history fundamentally repeats. There may always be an element of chaos but, notes Gray, 'historical perspective is the only protection against undue capture by the concerns and fashionable ideas of today',[17] a contextual position that suggests future warfare 'will be strategic history much as usual' regardless of weapons and other developments. It is Gray's position that disproportionately informs this primer and chimes with Clausewitz's suggestion that 'very few of the new manifestations in war can be ascribed to new inventions or new departures in ideas. They result mainly from the transformation of society and new political conditions.'[18]

A verso to that argument (and, in considering norms, yet another position on the argument's continuum) is provided in part by Strachan and the position that 'history is the study of change' and that students of transformation should instead embrace Liddell Hart's observation that 'the past is a foreign place' where situations never exactly repeat and, instead, it is change that is the norm.[19] Strachan is echoing historian Bloch whereby examination of 'how and why yesterday differed from the day before... can reach conclusions which will enable it to foresee how tomorrow will differ from yesterday'.[20] The position mirrors the instability of norms and behaviours described above and usefully factors for the changes in battlecraft's practice that are so clearly observable over recent years.

At its edge, the continuum is then occupied by McFate's notion of *Durable Disorder*, periods of deep, foundational unrest that are brought about from a new meld of long-dated and *systemically* unsettling factors; the resurgence of Russia; the rise of China and America's seeming retreat from the world's stage; from criminality, climate change and dwindling natural resources.[21] While all of these factors are likely systemic (and therefore important in norm movement), the issue for this primer is more the duration, depth and pace of these changes and the degree of change that they may have on battlefield behaviour. Finally, to this point, the argument's furthest reaches are occupied by the futurists and views around what they

[16] Strategic Studies Institute, *'Defining War for the 21st Century'*, SSI Annual Strategy Conference Report, 2010.
[17] Colin Gray, *Another Bloody Century*, Phoenix Books, 2005, 13.
[18] Strategic Studies Institute, *'Defining War for the 21st Century'*, SSI Annual Strategy Conference Report, 2010, 21.
[19] Hew Strachan and Scheipers Sibylle (eds.), *The Changing Character of War*, Oxford University Press, 2011.
[20] Ibid, 7.
[21] See https://www.seanmcfate.com/the-new-rules-of-war.

contend will be science's game-changing influence (artificial intelligence, robotics, machine learning) on the modern battlefield.[22] The reasoning that underpins these various positions is discussed in later chapters.

Battlecraft (and, by extension, its family of norms) is therefore based upon a whole family of intermeshing positions with their examination being complicated by non-obvious overlap across these themes. These messy intersections, moreover, are not helped by the dislocation between 'new' thinking around war's practice (the hybrid, asymmetric and irregular means discussed below), the degree to which these developments account for passing norms and, third, the current means and methods of undertaking the fight that are already available to commanders on the ground. Indeed, a challenge in considering revolutionary technologies is to ignore the advantages of *currently* deployed and otherwise non-disruptive technologies. Examples here include states' modernisation programmes underway within current arsenals. Most new military technology is not, after all, revolutionary. The tendency is for such programmes to be incremental to prior art with, therefore, correspondingly little impact on that continuum and, by extension, on conflict's passing behaviours. The long-dated norm, after all, is that new technology does not always signpost obsolescence in current weaponry.[23]

For the purposes of this introduction, it is also the *width* of commentators' framing which complicates attribution. All manner of adjustments get justified by all manner of drivers. Unsupported statements on war's practice highlight the subject's complexity, in particular around developments in battlecraft and how these might play out in their effect on passing norms. Analysis of war's forms and norms does not lend itself to abridgement or inappropriately condensed argument. The factors that drive norm change are not reducible to general statements on revolutionary technologies or their strategic effects.[24] Indeed, technophilia (here, the love of new battlefield means) generally undermines the premise that it is *how* you fight and not the technology *with which* you fight that is the enduring start point for conflict's norms. The primer's prevailing view is unsurprisingly that human endeavour (and not the tools with which battle is undertaken) remains the best indicator of success and that it is always the

---

[22] PW Singer et al., *Wired for War: The Robotics Revolution and Conflict in the 21st Century*, Mariner Books, 2015.

[23] Dima Adamsky, *The Culture of Military Innovation*, Stanford University Press, 2010.

[24] W Murray and R Millett, 'Military Innovation in the Interwar Period', *Cambridge University Press*, 1996.

excellent soldier who is best placed to neutralise the adversary's innovation, to capitalise upon their weaknesses and to win the war.

A further building block must be that long-dated advantage in battlecraft generally arises out of the deployment of non-revolutionary technologies. The long-lasting adoption of new means, after all, is dependent upon the overcoming of several deployment challenges and is often to ignore how 'low-tech' (and usually already deployed) alternatives can complement the integration of new practices.[25] In this vein, a passing norm remains that adaption of existing weaponry and its uses often produces similarly systemic transformation but does this through incremental and less risky adjustment rather than wholesale innovation.[26]

Nevertheless, while the introduction of disruptive weaponry may require material step change in hardware and software (and, as discussed in later chapters, 'humanware'), once deployed it may certainly posit a significant change in how humans wage war as well as the norms that will then govern its undertaking. Indeed, the recent pace in the development of these technologies appears extraordinary and it may therefore appear surprising that this primer devotes much of its time highlighting fundamental fault lines that must first be overcome before these developments can be introduced to the extent that their battlefield effect translates to changes in passing norms. These include the arrival of several missing pieces (in the case of autonomy, for instance, by how to code for ambiguity, how goals and aims and values can be embedded in machine routines, how algorithms can factor for data obsolescence and adversarial meddling) but also the solving of high-level actualities such as the interdependent and highly coupled nature of likely routines and, a recurring theme for this primer, the unsuitability of machine learning's current technical spine as the basis for weapons' independent operation. Given reduced supervision and dilution of meaningful human control in weapons' deployment, challenges around these technologies' statutory, moral and ethical ramifications must first be addressed if they are to be integrated in battlecraft.

Embedding new technology, moreover, has always required profound adjustment in the way parties organise their military efforts if the capabilities

---

[25] Robotics, Artificial Intelligence (AI), Internet of Things (IoT), Augmented Reality (AR), Analytics, and Robotic Process Automation (RPA) are some of the 4IR technologies.

[26] Nina Ann Kollars, 'By the Seat of Their Pants: Military Technological Adaptation in War', 2012 Ohio State University PhD text, http://rave.ohiolink.edu/etdc/view?acc_num=osu1341314153.

of new means are to be realised and this is unlikely to be any different going forward. And, as a lead to this introduction before considering the role of context in norm change, a final heuristic in the deployment of potentially disruptive practices (and thus consequential to the flexing of norms) is, of course, that the introduction of new battlefield technology rarely gives lasting advantage. Here, Boot's notion of nullification, the use of quite simple solutions to annul technical advances, echoes Gray's corollary that 'the history of war is not primarily the history of weaponry but instead of the history of the person who wields that weapon'.[27]

---

[27] Colin Gray, *Another Bloody Century* (London, Phoenix Books, 2005), 61.

1
Context's Continuum
Traditionalists, Pragmatists and Futurists

Changes in warfare's norms invariably arrive from a *cumulative* set of circumstances. Disruptions arise from a complex meld of trends, technical and legal and societal. They mirror adjustments taking place in politics, in procurement and current doctrine, in the ebb and flow of national interests, and what adversaries can do to upend these. Norm changes are always empirical, observable in time on the ground, and often driven by real politik and opportunity. Given this patchwork nature, it is therefore context that provides the prism which can appropriately weight each component in the debate. Here, while Russia's invasion of Ukraine certainly destabilises current norms and begs new analysis, it is unlikely that any *single* characteristic of that campaign will occasion wholesale changes in behaviours.

Instead, individual innovations may herald changes in a specific facet of battlecraft but, generally, it is their *aggregate* influence that occasions change. Attention, for instance, has undoubtedly been grabbed by President Putin's dependence upon brutal and conventional force, but norm change will empirically be determined by drawing from the overall picture that is Ukraine, from the conflict's humanitarian crisis and, in a new age of social media and ubiquitous information, visible reminders of the disproportionate effects of warfare on the most vulnerable. This is context at work and informs all aspects of battlecraft, from the political (energy, for instance, and its drivers around access and supply side diversification, the issues of food security, and the race for critical materials, equipment and commodities) to the granular (changes to battlefield modus operandi and the adaption of tactics and means in response to shifting context). Indeed, norms here must severally be impacted as supply chains adjust, global technology standards separate, financial-system effects crystallise and defence spending rises.[1] Context therefore remains *the* important tool in providing confidence to this primer's observations. It affords both a common analytical foundation as well as continuity from which to consider the degree to which the recent past is that foreign place suggested above and where it may be change that is the universal norm in war's undertaking. The verso, of course, is that it is context which damps down our over-reaction to events and advances, and where historical perspective protects against kneejerk reaction and inappropriate modification to behaviours.

A purpose of this chapter is therefore to identify where context acts as a change agent to behaviours. Catalysts must be considered in light of their often destabilising ability to bring about general change rather than specific and attributable adjustment to particular norms. This is invariably an outcome of *strategic* context and depends, of course, upon any one observer's set of assumptions. It is also borne out by the book's primary evidence base and the cacophony of views that arose here in what might appear quite straightforward chapter titles, each soliciting quite different responses from those interviewed in the book's early processes. Opinions on the degree to which systemic change is taking place in a vertical is unsurprisingly determined by how those participants observe, process and then telegraph those supposed shifts in current battlecraft and, given the importance of context, the spaces that are adjacent to each specific development. None of this benefits from ready abstraction. Context does not arrive in readily discernible packets. In the case of the battlefield, it is confounded by dual-use applications in emerging technologies, by lags in adoption (that depend more on exogeneities such as geography, third party interests and other incentives) as well as commercial imperatives rather than upon textbook models of take-up and assimilation. The enabling effects occasioned by the Apple iPhone, for instance, are a case in point as well as technologies based around, inter alia, the Internet of Things. While

---

[1]   Kevin Buehler and others, 'War in Ukraine: Twelve Disruptions Changing the World', McKinsey & Company, 9 May 2022, https://www.mckinsey.com/business-functions/strategy-and-corporate-finance/our-insights/war-in-ukraine-twelve-disruptions-changing-the-world.

innovation has long brought about improvement in neighbouring spheres, context points to a developing norm here being instead the *pace* of that change. Indeed, it may be that long-dated trends have simply become more apparent in an age of data saturation. Here, the recent decade has seen a move from 'everyone-with-a-phone' to a much broader phenomenon of universal (and therefore destabilising) access, one that is powered by ubiquitous data and seeming transparency, and the contextual swings that these developments can fashion.

Velocity and tempo affect context. First, the pace of today's heralded discontinuities, together with their technical and behavioural consequences, make forecasting norms a more involved, perilous exercise. Second, this is complicated by a widespread *revolution in expectation* (also emphasised by this project's original evidence base) whereby those same expected changes can have disproportionate influence in shaping narratives regardless of any twists that must first be overcome before these new means can be deployed. Conflation amongst the public and policymakers alike of what is 'possible', what is 'reasonable' and then what is 'probable' complicates analysis. Indeed, it is exactly this characteristic (and, for the purposes of this primer, this *trend*) that should prompt caution in the making of far-reaching claims, being inappropriately optimistic and confusing what is actually practicable and deliverable in the cold light of technical advances.

It turns out, of course, that expectation should often be regarded with suspicion regardless of its popular, political or professional roots. Ahead of Russia's invasion of Ukraine, for instance, this project's early commentators almost unanimously focused upon the certain deployment of precise, costless warfare. This seemed perfectly reasonable. Prior to February 2022, new weapons and their capabilities touted a list of disruptive advantages, from remote engagement and force multiplication to pervasive sensors (leading to new battlefield understanding and new attack surfaces) and an end to blue casualties, all uncontested examples of progress in battlecraft and all directly arising from newly available forms of warfare. But this, of course, was to ignore the empirics and practicalities thrown up by Russia's activities in its neighbouring country. It is also to ignore frictions (adoption, doctrinal and technical challenges) that complicate deployment and muddy attribution. After all, the practical verso here is a litany of previously botched procurement, the inefficient integration of new weapon platforms and practices, as well as logistical and behavioural tensions that have combined to dilute subsequent outcomes. The trend may also be reflected in a new norm of 'war extravagance' arising from the increasing disconnect between expensive weapons engaging inappropriately insignificant targets, the construct of 'attacking birds with golden bullets'.[2]

General Norm Divergences

Another contextual dislocation arises from differences that exist between the West's purported openness and value adherence relative to that of its autocratic adversaries. That this observation was repeated by the majority of participants interviewed for this primer confirms its importance as a narrative regardless of available hard evidence of norm change in the matter. This East-West, democratic-authoritarian argument is well rehearsed and has several angles, from adversaries' digital authoritarianism and their perceived ability to leverage central planning in order to create strategic advantage to, generally, a proven portfolio of non-lethal actions that autocratic adversaries are happy to deploy but which fall shy of triggering military response. Again, the phenomenon has a clear verso. Seen, after all, from the perspective of those adversaries, the lessons learned are certainly very different and more likely to be influenced by the West's controversial history of intervention in exactly these types of regimes.

This dislocation (and its attendant context) is exacerbated, moreover, by mercantile and economic-strategic considerations.[3] Indeed, this theme of aggravated competition crops up time and again in the book's initial set of interviews and, while these strategies are not new, they do point to an emerging norm whereby the means to undertake (and attraction of that undertaking) low-cost, opportunist and deniable meddling is quickly expanding.[4] That same technology which can democratise parties' ability to weaponise data and information, to shape opinions, to undertake non-lethal measures and to do this now at speed is inexpensive relative to the traditional hardware that has long characterised the battlefield. What does not change, however, is that all such new forms still depend, of course, upon the wherewithal, temperament and calculi of each adversarial party, together the context in which conflict will be undertaken.

---

[2]   TX Hammes, 'The Future of Warfare: Small, many and smart versus few and exquisite', *War on the Rocks*, 16 July 2014, https://warontherocks.com/2014/07/the-future-of-warfare-small-many-smart-vs-few-exquisite/.
[3]   Economist editorial, 'Globalisation and autocracy are locked together. How much longer?', 19 March 2022, *Economist*, https://www.economist.com/finance-and-economics/2022/03/19/globalisation-and-autocracy-are-locked-together-for-how-much-longer.
[4]   See, generally, Chapter 4 (*How Will Conflict Be Waged?*).

Norms' Ever Blurring Boundaries

Mapping context's role in the travel of passing norms is also a frustrating exercise as it heaps ambiguity onto already imprecise notions of behaviour, procedures and conventions. This might have been an easier undertaking in eras of uniformed soldiery, parties operating under a declared state of war and with otherwise quite understood boundaries. But while decision-making in times of ambiguity has always been a trial, countering a new sophistication in how adversaries deliver non-lethal measures brings with it a whole new set of challenges, especially given leaders' inexperience in these methods, the erosion in the effectiveness of traditional diplomacy, a waning in the efficacy of diplomatic instruments as well as difficulties in calibrating appropriate countermeasures. Moreover, the wearing-away of stabilisation mechanisms might appear illogical and contradictory given, for instance, the range of restrictions now being imposed upon Russia in response to its actions in February 2022. But it is precisely the breadth of these actions that highlights the imprecision and challenge (matters of attribution, enforcement, monitoring and, principally, in tuning) faced by coalition partners and their available means to change Russia's behaviour, the more so given the conflicting narratives swirling around that conflict. Indeed, parties' actions around energy assets, the freezing of large parts of Moscow's pre-invasion foreign exchange reserves as well as their far-reaching programme of secondary sanctions all reaffirm the importance that has returned to traditional geo-political measures in response to one party's overt fracturing of previously stable norms.

While few factors repeat exactly across all the testimonies of those interviewed for this book, contextual pointers are to be found throughout this primer's original discussion material (and which should make this book's appendix that abridges these interviews such an interesting subsequent resource). One participant's focus on, say, state-sponsored criminal activities as a tool in competition is another's spotlight on the rise of private military companies and the importance of mercenary forces over the timeline of this primer. While it is difficult to be confidently relativist, a developing norm here is that inputs to what were previously somewhat stable equations around conflict, geo-politics and defence planning have certainly become more volatile. They have also become less predictable. Speed of action and opportunities for obfuscation make definition more difficult, compounded by parties now deciding their own thresholds for action in ever more dynamic, speculative and less formulaic manners.

Areas of coalescence in that primary evidence obviously exist, notably around the *widening* definition of war and an on-oing blurring of previously established definitions of rivalry, conflict and outright warfare. Given (again) that a large number of small adaptions can quickly fashion fundamental change, such flux increases actors' risk, especially in circumstances where a party's actions have traditionally been tied to doctrine and standard methods. The importance, therefore, to norms is that war's practices increasingly meld the use of military *and* non-military activities in what is a broadening toolkit of forms and means regardless of parties' conventional strength. This has various components. Whether it be the role of digitalisation in shaping information or the globalisation of trade and its ensuing social and cultural ties, it is again the cumulative effect of these factors that together amplify adversarial options in, for instance, the fields of financial disruption, the manipulation of markets and opportunities for exploitation in cyber, resources and strategic real estate.

Notwithstanding Russia's playbook of conventional force in Ukraine, the emerging norm here is that wars are increasingly fought and won *beyond* the traditional battlefield. It is certainly not just about artillery pieces. Add, then, the contextual drivers and it becomes clearer to see, for instance, that velocity of change and multiplicity of means are increasingly quickly combining to change behaviours, whether through adaptions in combatants' supply chains, realignments taking place across coalitions as well as much more buffer being incorporated into parties' arrangements. All of these developments impose changes in practice, each providing new opportunity for misinterpretation and instability. In this vein, a further common theme has been participants' agreement around the erosion of human supervision in lethal engagements. This also has overarching contextual consequences as it must further reduce safety margins across battlefield activities and, in so doing, require command chains to make quicker, perhaps less informed decisions. Adjacent to this point, the Ukraine conflict demonstrates the disproportionate effects of these changes upon civilian groups.[5]

Volatility and challenges around attribution naturally inform behaviours. They broaden available means, whether from actors' 'vertical escalation' (an increasing intensity of action within given locations or in

---

[5]   Office of High Commissioner for Human Rights, 'Plight of Civilians in Ukraine', *United Nations Press Briefing Notes*, 10 May 2022, https://www.ohchr.org/en/press-briefing-notes/2022/05/plight-civilians-ukraine.

categories of competition) or from 'horizontal escalation' (the expansion of parties' actions, whether in new geographies, categories, or action types).[6] Volatility also arises from 'horizontal manoeuvre', the creation of bandwidth challenge from multiple adversarial activities being initiated across a wide and often unexpected set of disputes. While these characteristics may not themselves constitute new norms, adversaries' accelerating ability to leverage these practices serves to add to war's forms and, over time, its norms. Here, actors can rapidly price adversaries out of competing in a particular geography or in a particular vertical. An example here might be low-cost seizure of remotely located assets requiring the defender to commit to allocation of scarce defensive weaponry to protect those assets.

Unhelpful labelling, however, of *all* adversarial activity as 'warfare' contextually complicates both attribution as well as the fixing of norms themselves. First is the notion that militaries are the solution to all state-level problems. Second is that accelerating erosion of boundaries between traditional statecraft (albeit packaged for the digital era) and other forms of competition. Examples might include adversaries' use of blockade as well as cyber activities to steal and gain leverage in commercial domains.[7] Given that dividing lines between the military and commercial spheres have long been eroded, context is now just as shaped by actors' exploitation of dual-use technology as it is by as the collapsing cost of that technology across all parts of parties' supply chains.[8]

An adjunct behaviour then arises. Just as public expectations may now be framed by a consensus that 'everything' is now technically feasible, the norm change here is actually fuelled by a newly wide availability in 'good-enough solutions'.[9] While low-tech solutions can readily blunt high-tech means, norm change is taking place because previously non-peer entities may now reach near-peer capability faster, cheaper and (given new combinations of action) in perhaps unorthodox and therefore unpredictable manners.

---

[6]   For discussion, generally, on conflict escalation see: Forrest Morgan and others, *Dangerous Thresholds: Managing Escalation in the Twenty First Century*, RAND, 2008, https://www.jstor.org/stable/10.7249/mg614af.

[7]   Ken McCallum, 'Threat to UK from Hostile States Could Be as Bad as Terrorism, Says MI5 Chief', *Guardian*, 14 July 2021, https://www.theguardian.com/uk-news/2021/jul/14/public-should-be-alert-to-threat-from-china-and-russia-says-mi5-chief.

[8]   See, generally, Margaret Kosal (ed), *Proliferation of Weapons and Dual Use Technologies; Diplomatic, Information, Military and Economic Approaches*, Springer Cham, 2021.

[9]   David Vergun, 'Near Peer Threats at Highest Point since the Cold War, DoD Officer says', *DOD News*, 10 March 2020, https://www.defense.gov/News/News-Stories/Article/Article/2107397/near-peer-threats-at-highest-point-since-cold-war-dod-official-says/.

Resilience and Norm Change

A contextual observation is also that norm change is accelerated by parties' responses to their own or others' crises, the notion of 'lessons from other people's wars' and how this then affects parties' preparedness and resilience.[10] Empirically, this results in unexpectedly quick adjustments in practices. It also occasions parties' acquisition of leap-frogging capabilities from that adaption with, of course, ramifications then for escalation and a subsequent arms race in those new strengths. Norm developments also arise from parties' observation of *likely* adversaries. Identification of force discrepancies may be an imprecise and often political art but is also an oft-seen catalyst to move parties towards 'pacing threat', the matching of a party's benchmarks to adversaries whether through setting new targets or from allocating resources to exploit that actor's presumed vulnerabilities. While technology's ability to leverage such discrepancies may be attractive to parties, it is also likely to be broadly destabilising to passing conventions.

Norms, of course, morph in line with actors' own resilience in absorbing strikes and, in cases where there is clear mismatch in relative capabilities, the increasing means with which the lesser party may frustrate its stronger adversary's efforts, either through denying its ability to bring force to bear or, more likely, from the deterrence value in inflicting sufficient damage regardless of that party's more considerable and stronger means. Russia's experience in Ukraine provides a case in point and has been characterised by a meld of these three factors. First, the conflict highlights an increasing divergence between, on the one hand, large-scale combat operations and attendant high intensity war and, on the other, asymmetric means of fighting war that is designed instead to degrade, demoralise and attrite that supposedly stronger adversary, raising that party's costs of initiating its first encounter. The two conflict types require different mindsets, different structures and, as appropriate, tailored doctrine. Indeed, how a country fights remains more important than what it fights with.[11] Second, the enduring norm remains that the party must also continue the fight, even as its sophisticated means of doing so are destroyed or depleted. This truism also directs how allocators should deploy high-value weaponry avoiding,

---

[10]   Anthony Cordesman, 'Learning the Right Lessons from the Afghan War', CSIS, 7 September 2021, https://www.csis.org/analysis/learning-right-lessons-afghan-war.

[11]   Conrad Crane, 'Too Fragile to Fight: Could the US Military Withstand a War of Attrition?', *War on the Rocks*, 9 May 2022, https://warontherocks.com/2022/05/too-fragile-to-fight-could-the-u-s-military-withstand-a-war-of-attrition.

for instance, expensive missiles taking out low-value and irrelevant targets. As an adjunct norm and explored in later chapters, Ukraine reminds parties of the imperative of regenerating combat power and ensuring coordinated, recuperative capabilities.

It is interesting to consider the behavioural consequences of 'wrong' thinking on norms, whether this arises from optimistic assumptions being proved wrong, from sophisticated weapon platforms underperforming or when armies are blindsided by another case of technological surprise. A cure for failure is, of course, resilience, but this too is a complex, multi-faceted commodity and one that is particularly dependent upon passing context, parties' embedded culture and the degree of preparedness that each party's leadership has been able to instil in its wider society. Several frictions and contradictory priorities exist regarding the processes that comprise a party's resilience. Simple supply-side constraints are just one problem with which commanders and their procurement executive must contend. At the beginning of 2023, after all, it was generally estimated that it will take more than four years to replace just the Javelin missiles sent to Ukraine at current rates of production. Indeed, delivery time in late 2023 for 'new' generation weapons is generally reckoned to be more than 30 months.[12] Notwithstanding these platforms' success in the opening phases of the Ukraine conflict, the issue is best illustrated by Washington not having purchased any new Stingers since 2003. Also in this vein, the Arab-Israeli war of 1973 may only have lasted 19 days, but it took multiple years to restock weaponry to prior levels. The Ukraine experience must therefore shift norms in this regard. Here, anecdotal evidence suggests that it already refocused planners on the conversion of factories from producing the likes of household appliances to battlefield materiel. The norm is actually much wider. In considering resilience, today's high-technology platforms just take too long to build. International supply chains complicate this further, the aim of Western sanctions preventing technology transfer to Russia being to remove key weapon components from commercial circulation and preventing timely resupply.

Resilience requires raising the quality, availability, readiness and persistence of a party's means and doing this in a timely manner ahead of crises. The enduring norm here, after all, is that gearing up and mobilising

---

[12] Financial Times Editorial Board, 'NATO's Weapon Stockpiles Need Urgent Replenishment', *Financial Times*, 31 January 2023, https://www.ft.com/content/55b7ba35-6beb-4775-a97b-4e34d8294438.

for prolonged combat operations still requires considerable and often 'unexciting' investment. It requires that a party's organisational heft and the institutions delivering this effort are supported during years of peacetime and not be allowed to wither. In the case of the UK, a repeated accusation is that successive administrations have diluted these facilities, preferring instead the embrace of technology. Battlefield resilience, moreover, has several facets that range, for instance, from appropriately located and appropriately sized stockpiling of relevant materiel to, on the other hand, suitable casualty evacuation and the effective handling of parties' wounded, a capability that has often been relegated to out-of-theatre alternatives (that by extension must assume uncontested airspace) in order to ensure quick and reliable extraction. Apt treatment of casualties is a topic for later chapters but, as a norm, also influences mechanisms for those who are wounded to return to duty, a key contributor to the replacement of human resources in past conflicts. Indeed, a reinforced norm arising from Russian shortcomings in Ukraine underlines parties' overarching requirement to plan for lengthy, trying and extensive war.

Resilience norms must also factor for adjunct developments taking place *across* battlespace: artillery barrages displacing populations, bombing that creates refugees and brutality that foments political discord. Norms here must also take account of behavioural considerations, the consequences for instance of conscription, draft and parties' mobilisation plans aimed at shoring up deficiencies in human resources. How leaders manage their populace's expectations is a key constituent in shaping official narratives, maintaining calm and ensuring societal cohesion. Here, civilians removed from their homes and communities create tension as well as become a burden for the responsible government. Refugees are rightfully desperate and usually impoverished. As a category, they are also often weaponised, regularly exploited, and have very little leverage or voice to improve their lot. Unable to rebuild their lives, they challenge the bonds that tie together a society and, in so doing, tend to dilute efforts by the state to ensure resilience and manage dwindling resources.[13]

---

[13] Anne Applebaum, 'Russia's War Against Ukraine Has Turned into Terrorism', *The Atlantic*, 13 July 2022, https://www.theatlantic.com/ideas/archive/2022/07/russia-war-crimes-terrorism-definition/670500.

Context and Friction in Passing Norms

Context also provides a useful *inverse* to our understanding of norms and how they might move. There are, after all, counter-factuals to most accepted conventions and these readily (and regularly) conspire to throw grit into the patterns of behaviours that parties' best scenario planning is predicting from the set of circumstances immediately in front of them. Deployment challenges, for instance, as well as parties' patchy experience of integrating technology into battlefield processes usually accelerate such grit, the more so with second order effects then leading to further unexpected outcomes. Unlike doctrine, norms are not about providing didactic instruction. They generally reflect rather than teach. And they are regularly thrown into what might appear as disarray. A case in point is provided by the UK's ongoing challenges to replace its fleet of armoured fighting vehicles with General Dynamics' Ajax platform. Intended to provide a state-of-the-art reconnaissance asset for the British Army, the programme dates back to 2010 but is still unlikely to be deployed anytime soon.

The episode is not alone in destabilising wider procurement norms and, all at once, the new context is that parties should be dissuaded from acquiring untried technology. None of this is surprising. Indeed, counter-factuals and flip sides are themselves sources of inertia and become important components in the mix against wholesale change.[14] The role, after all, of the war planner is not to be too wrong and this weighs against either sweeping shifts in practice or doubling down upon untried methods. In considering context and its role in norms, change is always a long-dated commodity. Extended periods of relative stability generally damp down parties' calculations over existential risk and levels of threat to their own homeland and interests. The passing norm is that this in turn empirically waters down the willingness of those parties to undertake necessary systemic change.

Context then becomes key to navigating normative behaviour in an environment of conflicting agendas, of funding twists and behavioural plays that might otherwise suggest 'war as usual'. Faced with adopting a policy where available options are unattractive and divergent, the metric is that leaders rely on context and then are judged by history. Rather

---

[14] David Chinn and John Dowdy, 'Five Principles to Manage Change in the Military', *McKinsey*, 1 December 2014, https://www.mckinsey.com/industries/public-and-social-sector/our-insights/five-principles-to-manage-change-in-the-military.

frustratingly, norms then reflect over time the empirics of what happened. And leaders' choices are rarely, of course, clear-cut. On one hand, incrementalism in policy, procurement and doctrine can be as prescient as the spilling of blood in influencing battlefield processes. On the other is the promise, often untested, of new technology that convinces politicians and their procurement executive to invest again and again in new forms of warfare. It is this seesaw that creates yet another continuum in leaders' decision-making: Are chosen policies active or passive, interventionist or non-participatory, conservative or quite radical? In considering each norm, this continuum must then be overlaid with all of the constraints that are particular to that decision set, such as the rising specialisation of defence, the ever-shifting equation between human resources, the close and far fights, deployment decisions on trophy and hard-to-replace weapon platforms, and, as demonstrated by Ukraine's defence forces, the immediate effects that decision-makers expect from their available asset mix on any given battlefield.[15]

To this point, a new norm of conflict then becomes the goal of *minimum* deterrence that is credible in each specific environment. In order to do this, norms must factor for the widest set of war's available forms. Prior to February 2022, this may have led parties to field smaller but higher-end fighting forces in a portfolio of means that 'fall just short of a war' war. That course of action may still be valid (a mix of means that includes hybrid and conventional means) but, following Russia's invasion of Ukraine, the context of conflict into which war's forms and norms must be sited has again changed and, for the UK, behaviours must now factor for the building of resilience, the shifting of force postures in light of the empirics informed by Ukraine's battlespace, prioritising the protection of the country's centre of gravity (alliance cohesion) as well as greater emphasis upon thematic communities of interests. Unsurprisingly, outcomes here will take decades to play out.

The Expanding Nature of Available Means

Nevertheless, the primer's overarching context returns again and again to the expanding nature of available means, itself an emerging norm of

---

[15] Gaston Dubois, 'MANPADs in Ukraine: the return of Russian aircraft's biggest fear', *Aviacioneline*, 7 March 2022, https://www.aviacionline.com/2022/03/manpads-in-ukraine-the-return-of-russian-aircrafts-biggest-fear/.

warfare by which adversaries have a growing portfolio of tools with which to achieve strategic aims. First, the book's evidence reiterates the increasing difficulty of measuring deterrence and its utility. Warfare is getting sneakier with a quickening dislocation between the three components of capacity, will and signalling. Here, interviewees pointed oftentimes to the UK being 'in the business of deterrence rather than the business of defence'. The evolving norm might therefore be that deterrence is increasingly (and unhelpfully) bespoke and particular to individual adversaries and scenarios.

This is a departure from earlier eras when enough ships and enough land forces, properly positioned, victualled and led, were together sufficient to deter all of a party's adversaries and, it was reasoned, preserve that party's set of advantages. While changes to this equation have long had consequences upon an adversary's strategic calculus, the extrapolation here has been (and remains) that political will and a party's embedded determination to fight form the overarching contextual imperative. Indeed, this will continue to be the case notwithstanding what might appear to be the norm's eclipse in the recent din around relative capabilities (and, incidentally, the instability in deterrence that this often entails). Russia's opening campaign in Ukraine gives weight to this observation. It may be an unsatisfactory generality but it is useful to highlight this increasing frailty in previously assumed relationships, the interconnected nature of behaviours in conflict, the volatility of party's priorities and, as a result, the wide dispersion of outcomes (here, changes in behaviours) that seemingly minor changes in battlecraft and posture can drive. Indeed, context is indispensable precisely because it shows up again and again the poor abstraction of inputs that constitute norms, a useful characteristic to bear in mind when delving deeper in the coming chapters.

2
Information and the New Importance of Data

Wikipedia reckons that since 2021 there have been more than 100 wars. Understanding this number and its segmentation provides useful insight into current forms and norms of warfare given the sheer frequency of these events. Their classification, after all, covers wars of ethnicity, of counter-insurgency and of genocide. It includes drug wars and wars for the purposes of the state, wars of identity and wars of independence. In considering changes in normative behaviour, an important pointer to current practice is the proliferation of labels now used to describe these wars and warfare. While not individually new, they do reflect a nervousness around future wars' shape and how it might be shifting quicker than parties' capacity to process and prepare for it. The pantheon of terms now used to discuss war speaks increasingly to its technical processes ('fourth generation', 'fifth generation', 'algorithmic'), its means of undertaking ('full-spectrum', 'mosaic', 'hybrid', 'asymmetric'), its degree of conviction ('liminal', 'grey zone' and 'shadow') as well as its means of prosecution ('vicarious', 'proxy' and 'surrogate'). Voguish terms relate to its pace ('accelerated' or 'endless'), its genesis ('economic', 'preventive', 'imperial' and 'identity') and its execution ('manoeuvre', 'offensive', 'defensive').[1]

---

[1]   Much of this chapter's introduction is informed by Matthew Ford and Andrew Hoskins, *Radical War: Data, Attention and Control in the 21st Century*, Oxford University Press, 2022.

Two observations arise. First, while individually not new, it is the *array* of terms and the increased commonality in use (both individually but also in combination) that suggest change is in train. In so doing, it also confirms the opacity of activities that comprise modern warfighting. A second reflection is that current language remains frustratingly ill-suited to discussions and definitions around national security. It alienates politicians, soldiers and public alike. The norm here has long been that audiences are generally disadvantaged by impenetrable concepts and acronyms, the more so given that norms can appear reductive and trite with a tendency to obscure the horrors of war. It is not, of course, the intention of this chapter to minimise these matters but there is, after all, a recognition that the term 'effects' has been diluted almost to the point of formlessness. Indeed, it is important throughout this primer to remember that analysis must always distil down to war's very nature, the killing and destruction of the adversary, the horror and butchery that make up the unchanging basis of warfare.

This plethora of terms also mirrors the erosion in binary classifications between peace and war, combatant and civilian, battleground and areas of uncontested calm. No longer can conclusions be drawn simply from the mere 'size' of a conflict, the weapons used in that conflict or the means with which political aims have been pursued. Instead, it is necessary to look into parties' societies and conventions in order to establish the tipping points and discontinuities that will inform passing norms. To this end (and the focus of this chapter), the book's authors consider recent developments in the world's *digital* practices to be sufficiently discontinuous and sufficiently material to merit their own chapter in an effort to establish variation from prior art and in order to fix how norms might flex over the primer's period of interest.

This phenomenon has several angles that require examination, the aim of this chapter. According to agencies of the United Nations, more than half of the world's population was already using the web by 2019 with more than four billion consumers of online information and services able to communicate digitally with their fellow citizens. According to Internet Live Stats, this figure has risen by nearly a third in the intervening two years to more than five billion users. This trajectory is extraordinary. As of 2020, 96 per cent of the world's population had access to a mobile-cellular telephone network.[2] Similarly instructive to future norms is to look at the hardware enabling this explosion in connectivity. Statista estimates that by

---

[2]   Internet Live Stats, see https://www.internetlivestats.com/.

2021 there were 6.3 billion smartphone users in the world, a 70 per cent increase in users from just five years previously. More than four million smartphones are sold every day. Connectivity is nowadays *all* about handheld devices with less than 12 per cent of the world's population now connecting to the Internet by fixed-line telecoms.

Connectivity and Norms

The numbers of connected users and the behavioural ramifications of the smartphone as a social and information node have upended how war is chronicled, understood and amplified. This extraordinary scale of adoption (both the number of handsets but, more importantly, the services, interventions and scope of capabilities that the smartphone enables) is an important factor to this primer's deduction on how systemically adopted changes in behaviour, in this case personal cellphones and their ecosystems, will obviously inform battlefield conventions over the period. The revolution here is twofold; one the one hand, it is about the immediate availability of information but, on the other, it also concerns that information's authority, reliability and means of verification. It is about veracity and trustworthiness.

News, for instance, from the Ukrainian conflict is immediately published, considered and shared on platforms such as YouTube, Facebook and Twitter (now 'X') in a manner and speed previously considered impossible. Feeds now provide the world's population with a frictionless ability to see, judge and engage, and to do this within a constant barrage of information and data points that pass as fact. This in turn underpins a whole series of new behaviours whose effects cannot be underestimated with users now able to move seamlessly from the phone in their hands 'to their minds, to their memory and, as it is for some, instantly forgotten with a simple swipe of the screen'.[3] It points to an expansion of space that war can now reach even if battlefields may retain their traditional physical, geographical location.

A norm change for this primer is that it is increasingly difficult to be a bystander in war. As people use their devices to record their surroundings and experiences, they may unwittingly be transmitting data points that are useful, for instance, to generating battlefield targets. As noted by Ford and Hoskins, this very act of participation 'collapses the boundary

---

[3]   Matthew Ford and Andrew Hoskins, *Radical War: Data, Attention and Control in the 21st Century*, Oxford University Press, 2022, xix.

between those who observe war and those who engage in it, lulling actors into a false sense of being active, of making a difference, creating shaky expectations that information translates into both knowledge and action.'[4] The new norm here is that a whole set of war's traditional parameters are being reconfigured around a quite different conception, the *digital* individual. The gravitas and belief in broadcasters sitting behind their desks and reading from a prepared autocue have been replaced by billions of users and user-created narratives. A host of adjunct changes in practice have accelerated the trend. Newspapers stumble over legal constraints and editorial processes, over corroboration and fact-checking. At the same time, re-posting and an explosion in social media apps act as multipliers for this personally collected and curated but essentially unverified commentary.

Such reflexive posting also means that previous distinctions between civilian and combatant, a central tenet of the laws of armed combat, are increasingly blurred by the scale and reach of social media whose practices have irrevocably extended the battlefield beyond its traditionally physical space into a multitude of different contexts and agendas, all without boundaries or oversight. Every global citizen can now produce, publish and consume media all from one device regardless of their proximity to the fighting. And this is accelerated by new networks and limitless public platforms that enable anyone within or outside of a conflict zone to tell their story, expose events, involve themselves and participate in informational warfare. The empirics of this, moreover, is both to dehumanise the act of warfighting while also upending the process of learning lessons. A developing new norm is that each military event now produces multiple narratives, twists and storylines, usually uncensored and all emerging at different speeds and to different timelines. Second order effects are legion, unpredictable and likely unfolding at a very different tempo to military activities on the battlefield.

Social Media as a Change Agent in Norms

Smartphones are therefore the vehicle through which an inestimable number of users can participate in conflict in ways that were even recently considered inconceivable. This embedding is doubly powerful given individuals' reliance upon their devices and networks for all manner of everyday tasks. The phone is a bank teller, an authenticator for the user's security, and usually the sole engine of written and verbal communication

---

[4]   Ibid, 47.

for the phone's owner. It is a diary, a journal and the platform that the user relies upon to capture, aggregate and then disseminate all of that user's experiences. It is the new glue between families, friends, societies and interest groups. The commercial imperative for social media assets, moreover, is to keep these digital individuals online whether by sustaining the pursuit for 'likes', by pushing like-minded information and by creating apparent connection to others in a web of perceived tribes: 'these circumstances create', notes Ford and colleagues, 'the conditions for a connective turn that the entire epistemological framework for understanding is deeply mediated'[5] and, in so doing, the norm of encouraging spectacle over rigour is reinforced, likely flattening appreciation for context with information being transient, volatile and often only fleetingly considered by recipients.

Recent conflicts evidence the smartphone's new influence upon behaviours. First, technological innovation has helped Ukraine to offset certain of Russia's conventional advantages by blurring the lines between civilian and military actors and doing so to an extent not previously seen in combat zones. This certainly constitutes a developing norm in warfare given the new empiric of citizen involvement in digital warfighting.[6] A second insight arises from the inability of international institutions and legal frameworks to keep pace with digital developments. Content providers operate largely without constraints and, unsurprisingly, it is increasingly fuzzy how rules of law apply in a digital conflict. A third trend has then been the *quantity* of information produced, the uses of this data and how this data is subsequently stored and retrieved, much of which might, after all, be potentially useful in establishing accountability in engagements post facto. These traits constitute disruptive and newly permanent characteristics of war that require flex in norms. Information technology is now *the* central feature of warfare, requiring new policy, changes to doctrine and adjustment to the shaping of assets towards war's prosecution.

While much coverage is rightly focused on the impact of kinetic weapons in Ukraine, digital assets such as Space X's satellite-based Starlink services have had a disproportionately important impact on battlefield practices evidenced, for instance, in new efficiencies and means of coordination, upended reconnaissance and surveillance, better

---

[5]   Ibid, 75.
[6]   Steven Feldstein, 'Disentangling the Digital Battlefield: How the Internet has Changed War', *War on the Rocks*, 7 December 2022, https://warontherocks.com/2022/12/disentangling-the-digital-battlefield-how-the-internet-has-changed-war.

artillery allocation and the deployment of crewless assets. Services such as Starlink have also enabled Ukrainian citizens to involve themselves in the engagement of invading forces, extending warfighting beyond traditional military and government actors. While intelligence from civilian sources has long been a feature in military operations, the new scope afforded by digital developments is itself a new and permanent norm of warfare. In the case of Ukraine, this has been facilitated by its government's deployment of crowdsourcing apps giving individuals the means to upload critical information on Russian military movements and assets with citizens now able to submit geolocated photos and evidence of Russian military sightings through a commonly available phone app.[7] Resulting data is then aggregated and used by intelligence officials to determine and prioritise targeting. Going forward, such operations will continue to help governments galvanise international support to maintain the flow of arms and other forms of aid.

It is also the new *scale* of digital operations on the battlefield that constitutes a discontinuity. In this vein, Ukraine's defence ministry was quick to coordinate closely with a self-styled 'IT Army' in its efforts to target Russian infrastructure and websites, a cohort composed of some 400,000 volunteer hackers.[8] More than 1,000 civilian drone operators then contributed to Ukraine's defence by surveilling Russian assets from the air and relaying that information to Ukrainian military units for artillery strikes. The verso exists here for norms in that these developments pose difficult questions regarding civilian protection under international humanitarian law, especially in its bedrock legal concept of distinction whereby parties delivering lethality are required to distinguish between civilian population and military personnel, directing operations only against military objectives. They also impact on day-to-day practices as evidenced by Google's decision early in the Ukraine war to disable certain live traffic features offered in its mapping services; while traffic patterns might serve as a useful information source for those fleeing hostilities through the signposting of possible exit routes, the technology also enabled Russian forces to identify road circulation patterns and derive possible targeting opportunities from such data.

---

[7] Vera Bergengruen, 'How Ukraine Is Crowdsourcing Digital Evidence of War Crimes', *Time Magazine*, 18 April 2022, https://time.com/6166781/ukraine-crowdsourcing-war-crimes/.

[8] Jennifer Shore, 'Don't Underestimate Ukraine's Voluntary Hackers', *Foreign Policy*, 11 April 2022, https://foreignpolicy.com/2022/04/11/russia-cyberwarfare-us-ukraine-volunteer-hackers-it-army/.

This recent ubiquity of mobile connected devices therefore revolutionises how armed forces must now organise for warfighting, from new processes for the collecting of data points to the refinement of targeting and decision methods before engaging targets informed by that information. On the one hand, this feeds an expectation that it is now possible to identify targets long before applicable parties realise they are being targeted. On the other, Ukraine's actions in the summer of 2023 to retake territories, undertaken in conditions of concealment and secrecy, demonstrate that omnipresent news equates to neither insight nor understanding of intent. Warfare is becoming as much of a duel between fighters in combat as it is a complex cat-and-mouse informational exercise.[9]

Ramifications arising from, say, a single digital snapshot help us understand better the reach of social media on warfare's norms. That one data grab can spawn a portfolio of interesting detail, from target points generated from the connections, footprints and other associations derived from metadata to revealing patterns of life that may be derived from that image. It is the bounty of possible information, the available statistics and second order intelligence from that one data set that are set to be so disruptive to battlecraft. The observation, however, has a verso. As is the case so often in this primer, flip sides affect behaviours in unexpected (and often unattractive) ways. Here, data sets can spin information in a myriad of ways. Episodes can be manipulated and faked in a manner less conceivable with earlier analogue information. Indeed, just as the smartphone has facilitated a wholly new audience to war's innermost workings, those in a battlespace are now able to recast and transfigure exactly that same information, but in ways that secure advantage and highjack initiatives, and to do this in ways that are morphing very quickly.[10]

Gone are constraints upon individuals who gather, publish and consume media. This fundamentally disrupts parties' ability to promote and then monitor an 'official' narrative. The emerging norm then becomes that authorities must either co-opt commercial entities that manage these data flows or otherwise risk pushback in their attempts to regulate them in their effort still to dominate those narratives. This represents a disruptive

---

[9] See, generally, Chatham House Feature, 'Seven Ways Russia's War on Ukraine has Changed the World', *Chatham House Editorial*, 20 February 2023, https://www.chathamhouse. org/2023/02/seven-ways-russias-war-ukraine-has-changed-world.
[10] David Ignatius, 'How the Algorithm Tipped the Balance in Ukraine', *Washington Post*, 19 December 2022, https://www.washingtonpost.com/opinions/2022/12/19/palantir-algorithm-data-ukraine-war/.

transfer of power for which militaries must now factor. At risk, moreover, are all of the foundations upon which force may be justified by a state player. And losing sway over this can prompt a significant rejoinder as evidenced by events in summer 2023 and the informational activities around the Wagner Group, a private military company. This is no easy task. It is, after all, the very abundance of platforms and media practices that can create this huge set of perspectives and narratives, creating a crisis of representation where consensus about war may suddenly become unstable in ways not previously experienced by chains of command.

Other technologies have similar disruptive potential. Blockchain, for instance, may not be a weapon system in itself but is an emerging enabling technology of the new information environment. At its most basic level, it provides a unique means to store data using cryptography to maintain a chronologically ordered record of transactions. The development also highlights emerging imbalances between potential adversaries and their real degrees of participation across the emerging technologies that are likely to shape operations over the coming two decades: While China filed nearly 33,000 patents in this field in the six years to 2021, Western parties lodged less than one third of this number.[11] Blockchain's impact may be difficult to assess (and its ultimate relevance remains a point of conjecture), but it provides a useful proxy to other developments now in research (quantum computing, next generation machine learning, machine sentience) and its reach is reflected today in several parties' investigation of the technology as a framework, for instance, upon which to issue their own digital fiat currency, a further disruptor to long-stable geo-political behaviour should such means become reality.

All of these digital practices, after all, may in future be weaponised and so influence tomorrow's norms. In the event, for instance, of a non-US central bank achieving widespread adoption of its own currency and systems, the current structure of economic sanction strategies might be upended. To this point, North Korea's theft of $250 million in crypto currency in 2018 is already an example of digitally enabled theft and, as a second order effect, of effective sanction busting, an emerging norm here being that parties' integrated deterrence may require profound rework. As an aside, other relevant models for blockchain and its proxies include

---

[11] Mike Knapp, 'The United States is Behind the Curve on Blockchain', *War on the Rocks*, 30 August 2022, https://warontherocks.com/2022/08/the-united-states-is-behind-the-curve-on-blockchain.

the technology's position in cybersecurity's zero-trust architecture and its ability to inventory software such that patching of machines can be undertaken reliably, an important component in the roll-out in due course of, for instance, autonomous weapons and the delegation to machines of decision-making around lethal engagements.

Data Ramifications on Norms

There are also particularly near-term consequences to norms from these developments. First, layers of the military's current processes and platforms may abruptly be redundant and require systemic reform. The smartphone's facilities, after all, make it possible 'to mobilise populations, replacing the rifle', notes Ford, 'as the weapon of choice for those engaged in mass participation in war.'[12] In this vein, mobile telephony and ubiquitous data availability has broken down previous hierarchies to a point where everyone now has a view, is a victim or perpetrator, and where distinctions between audience and actor, soldier and civilian, media and weapon have become manipulated or, at best, blurred. An emerging norm must therefore concern the *management* of public opinion, of different audiences where the population's will is now more segmented, polarised and volatile.

This is all exacerbated, of course, by the ability of online polemic to impact public sentiment very quickly and without check, the consequences of unverified opinions impacting both the boundaries and trigger-points of conflict. Polemic flattens experiences and saturates senses. At the very least, developments here require that politicians and those charged with executing their goals re-examine constantly how war's changing processes impact their many constituencies given what is now a much more deeply intermediated world. That current practices are less and less fit for purpose may be a long-dated truism but, while this may be a frustration for parties, the effect here is that all of these dislocations throw up opportunities for others to advance their own often opposing interests. And, of course, to mess up current arrangements.

Another ramification to norms that stems from this sudden ubiquity in public connectivity is the 'de-territorialisation' of conflict through the seizure of minds, opinions and influence. The capability is suddenly now easier and more available but also *cheaper*. Commanders' courses of actions must reflect

---

[12] Matthew Ford and Andrew Hoskins, *Radical War: Data, Attention and Control in the 21st Century*, Oxford University Press, 2022, 10.

this but also note certain new constraints that these developments impose. Time, for instance, is the new multiplier of data. During any one minute in 2020, Zoom hosted more than 200,000 people, YouTube added 500 hours of video and Instagram hosted nearly 350,000 stories. The discontinuity here is that connectivity saturates individuals with data such that it is ever harder for parties to make sense of what is in front of them. As again noted by Ford and Matthews, 'stories emerge asynchronously into our online worlds… New meanings can be constructed out of this new context… [breaking] and reframing experiences in ways that only serve to distort the what, when, why and who of war'.[13] The new norm is that parties may know more about everything but actually have less ability either to understand or to control the variables that make up these storylines.

The issue, of course, has a verso. Just as data overload does not lead to understanding, other frictions (disinformation, selection biases and adversarial intentions) combine to complicate provenance, acting as a dampener to profound norm change, the more so given the empiric of these platforms' patchy performance on the ground. Two contradictory observations arise. In the case of social media, users turn out to be extraordinarily susceptible to voguish fads. There is little loyalty in consumers to the platforms they use. Users, however, and how they live their lives, are very quickly reliant upon their devices and the information feeds they provide.

The effect is also felt across conflict's institutional circles, for instance within traditional broadcasting and their news readers caught between a panic over fake news and a wistfulness for how reportage used to be undertaken. Moreover, missteps by those news readers and by platforms alike serve to erode audiences' trust in what is seen and read. In terms of norms, therefore, the consequence of what promised to be a new age of digital openness may often be a dilution of consensus and the sowing of doubt and polarisation. Fragile connectivity (or, at worst, total denial of service) also has a disproportionate effect on those dependent on coverage in conflict zones. In this case, operating in a condition of blackout, all those same banking, security, communication and associated facilities provided by the smartphone (and which have upended and then reshaped daily life) are suddenly gone. Parties are denied the platform to control the narrative and, empirically in these circumstances, become more susceptible to alternative narratives.

---

[13] Ibid, 26.

The complexity for norms is that *all* of these developments are both informed by and then informing of war's prosecution. An uneducated forecast, moreover, must be that retail smart devices will continue to penetrate populaces and continue to interact with battlefield practices in ever more systemic ways that shape the nature of how parties see and understand their environment. Important for norms is that this ever-wider connectivity also creates more 'attack surfaces' for parties to act upon. Here, for instance, the freely downloadable WhatsApp has been used to marshal opinion, organise movements and even coordinate military attacks.[14] The smartphone has itself become a wide-ranging sensor, capturing data on targets' status, movement and location and leading commentators to muse that the traditional Clausevitzian Trinity and its model of war should now be recast to understand political violence across the three dimensions of data, attention and control, together the *datafication* of battle.[15] In this case, perception has become a newly available battleground complicating as never before the relationship between fact, knowledge and understanding.

A continuing norm of warfare remains the pursuit of information for military purposes. Military forces invest in intelligence capabilities to divine as much data on their situation as they are able on enemy dispositions, force structures, locations and, to the extent possible, intent. The accompanying norm, of course, is that parties now face a data management problem rather than a data collection problem. The new challenge is around data's delivery and manipulation in order for those parties' commanders to be able to make a decision at the 'speed of relevance'.[16] A corps commander, moreover, requires different information to the platoon commander. Here, the verso, is that information empirically *slows* decision-making given the data that must now be verified and process. This reinforces the importance of preparation before a conflict starts, providing an opportunity to work through large amounts of information without the time constraint of continuing operations. Pre-designating targets and created target packs in peacetime, or a more nuanced 'pre-conflict' period, will continue to be resources well spent.

---

[14] Inna Lazareva, 'WhatsApp Finds New Uses in Conflict Zones', *Reuters*, 3 August 2017, https://www.reuters.com/article/us-global-crisis-health-tech-idUSKBN1AJ0UX.

[15] See, generally, ICRC, 'Digital Technologies and War, Vol 102, 913, https://international-review.icrc.org/sites/default/files/reviews-pdf/2021-03/Digital-technologies-and-war-IRRC-No-913.pdf.

[16] Joe Dransfield, 'How Relevant is the Speed of Relevance?: Unity of Effort Towards Decision Superiority is Critical to Future U.S. Military Dominance', *The Strategy Bridge*, 13 January 2020.

The sharing and publication of data is another developing norm. Military intelligence used to be circulated only to vetted groups. Intelligence was considered sovereign and a key means to maintaining competitive advantage such as the intelligence sharing relationship underpinning the Five Eyes community of the US, UK, Canada, New Zealand and Australia. While the circulation of intelligence remains restricted, the war in Ukraine has seen states share information in the public domain in a way that has not been done before.[17] The British Ministry of Defence's daily Defence Intelligence update is an example of this, whereby military intelligence assessments are published on Twitter for all to see. This links to the social media norm discussed throughout this primer. While these updates may be fairly bland (and often to be found from open sources), parties' reported sharing of satellite imagery and information derived from clandestine sources is a noted departure from recent practices.

A further developing norm is the use of intelligence to call out disinformation and highlight atrocities and to do this in the public domain. The development also provides a useful prism to inform the general public at a time when significant resources are being funnelled into Ukraine notwithstanding domestic travails in those supporting countries. A verso here is that such intelligence sharing is regarded by adversaries as antagonistic action. A verso to the verso is that information sharing also serves to reveal the extent to which information that an adversary thought was secret is known by other parties. Interestingly, the practice predates the Ukrainian conflict with the US and UK briefing publicly on likely Russian intent.[18] While not yet a new norm, an important precedent has been set.

Data sharing is also an increasingly important practice if operations on the ground are to be optimised. The current practice, after all, is that military systems have been brought into service at different times, based upon different and usually disconnected systems. It remains a priority, from one generation of commanders to the next, that their hardware talks to each other, that sensors, deciders and effectors are seamlessly connected.[19] The aspiration, after all, is that a remotely piloted air system sends details of a target to another nation's headquarters where a determination is quickly

---

[17] Huw Dylan and Thomas Maguire, 'Why are governments sharing intelligence on the Ukraine war with the public and what are the risks?', *King's College London*, 27 September 2022.
[18] Julian Borger and Dan Sabbagh, 'US and UK trying to fend off Russian invasion by making intelligence public', *The Guardian*, 16 February 2022.
[19] Congressional Research Service, 'The Army's Project Convergence', *Congressional Research Service*, 2 June 2022.

made on its efficacy and legal status before being handed off to a third country's artillery battery. This remains difficult to do within a force, and even more challenging across nations. Nevertheless, connectivity of this sort is being demonstrated in Ukraine. An Android application called GIS Arta has reduced targeting times for Ukrainian forces from 20 minutes down to one.[20] Known as 'Uber for artillery', the app allows efficient matching of supply and demand. Targets can be registered from a range of inputs such as uncrewed aerial systems, smartphones and satellite imagery. The software assigns targets to the most appropriate weapon system that is in range and currently available for tasking and sends through a fire order. Military forces, however, remain tribal entities. A significant organisational change would be required to break away from traditional structures and the sharing of data must be a first principle in all endeavours if norm change is to be achieved. Military problem sets, after all, will otherwise continue to be viewed through the lenses of those domains and the respective services which are dedicated to them. Land wars are conducted by generals and naval battles by admirals.

Intelligence Sources

Another recurring theme from the primer's evidence base is that the battlefield effectiveness of smaller and less resourced forces has increased over the past decade relative to more established, better funded parties and that this trajectory is set to continue. It is easy to see why, given developments in the digital sphere: the cheap but clever meme that goes viral or the isolated incident that is captured on a smartphone's video and is downloaded by several hundred million users in the course of a month can have extraordinary influence on opinions and parties' actions. That this is now possible makes digital assets interesting to warring parties but also, of course, particularly susceptible to meddling. Indeed, whether this concerns the militaries of Russia or Iran or non-state parties intent on mischief, basing any analysis on long-dated practices that predate these digital developments seems increasingly irrelevant. It instead requires new tools, a new frame of mind and better understanding of the assets' capabilities as well, of course, as a new confidence to capitalise upon vulnerabilities.

This new assurance is already evident across battlecraft as evidenced by commentators' conviction that they are witnessing the same revolution

---

[20]   Charlie Parker, 'Uber-style technology helped Ukraine to destroy Russian battalion', *The Times*, 14 May 2022.

in intelligence that is akin to the 'revolution in expectation' around new technologies: 'I've seen it therefore it exists'. Certainly, norms must factor for almost daily advances in the capabilities of battlefield digitalisation and that same 'datification' that characterises engagement processes, from breakneck improvements in artificial intelligence, in data science and quantum computing and the certain disruption that this progress will drive in cheaper processing, better tools, improved machine interventions and the like.

Intelligence, after all, has been upended across its processes by new digital practices. Open-source intelligence has become the catch-all for what is a highly diverse form of data collection and analyses that is transforming parties' decision processes on the battlefield and, as such, occasioning an evolving norm of warfare. While questions may remain about the techniques' ongoing automation and, statistically, the reliability of its outcomes, the method's dependence upon publicly accessible sources of information (whether that be social media, academia, press reporting, government data or other unclassified resources) means that the tool can only ever be as effective as the sources that comprise its inputs. The norm, however, is that its utility continues to be enhanced by new software tools, improved processing and intelligence assessment as well as the layering of multiple data feeds to achieve better outcomes.

The extent to which this will cement new battlefield practices (and thence norms) will therefore depend upon that progress while still acting as a valuable adjunct to otherwise traditional intelligence processes. Indeed, developments currently underway in satellite imagery, radio frequency data, synthetic aperture radar and, broadly, across widespread commercial services all suggest that the reach (if not the accuracy) of battlefield intelligence will continue to develop disproportionately over the primer's timespan. Given that its contributory sources are rarely classified and therefore unusually accessible, the evolving norm must be that open-source intelligence will continue to mature as a tool that is quickly able to inform both public and military decision-making. That it is unlikely to replace human assets is less important than the additional clarity it offers to war's processes.

Narratives and Norms

A tangential norm of war revolves around the construct of 'internet infrastructure' and the implied permanence of digital data. Indeed, a further verso to intelligence's seemingly crisp practices is that the capture, analysis and subsequent retrieval of data from the online environment is actually a disorganised, frictionful process with few rules. Data is volatile and prone to decay, its management is sensitive to exogenous factors, to human error, whim and mismanagement, to deletion and broken links. This represents a surprising departure from pre-digital information practices where meticulous human methods, archiving and cross-referencing were in-built attributes to the intelligence process and subsequent analysis of information. Any such randomness in the handling of that data is significant, the more so given the revolution in data-to-news practices discussed above. Just as digital material now facilitates the open-source analysis and understanding of war, it also does this on the basis of a more fragile information ecology.

Emerging norms in this space must similarly factor for the same instability that characterises the internet's enormous pool of data, its opaque embedded misinformation and flawed associations. Just as open-source intelligence (OSINT) relies upon seamless receipt of information, so those undertaking complex intelligence (whether by hand or, the context of this primer, by machine) must understand the efficacy of data inputs that comprise their processes. Partial data, after all, requires expert backfilling and the layering of assumptions, a current source of weakness in unsupervised machine processing where out-of-the-loop humans are unable to step in to oversee outputs. Indeed, this human-machine balance (considered in detail in later chapters) is precisely important because of the danger, over time, that confidence is generally eroded in the accuracy and resolution of data that is held digitally. Most users, after all, neither understand nor manage the way their data is located, providing new attack surfaces, fuelling uncertainty and creating further opportunities for mischief. This is a whole new battleground and fosters an emerging norm for how conflict will be undertaken. Moreover, the simple volume of digital media denies *any* possible orderly organisation of that data.

Given that the archival facility of the internet hangs together on a largely random basis (absent of the creation, curation, ordering and seamless retrieval that parties might expect from a formal archive), this datafication will increasingly determine how 'war is perceived, experienced, won and not won, legitimised, declared, fought and lost, studied or ignored, hidden and made visible for different actors, for different ends, in and over time'.[21] This is a pivotal observation. Indeed, that online world is now completely

---

[21] Matthew Ford and Andrew Hoskins, *Radical War: Data, Attention and Control in the 21st Century*, Oxford University Press, 2022, 76.

'entangled with remembering and forgetting'. This has a difficult verso. Research suggests that YouTube alone removed nearly eight million videos in the period between July and September 2020, more than 90 per cent of which were taken down through automatic filtering. From a human rights perspective, much of this data would presumably have contained information essential for future evidence-gathering and prosecution.[22]

Changes in norms also arise from the transformation of narratives and how they are understood by each relevant actor. Given the digital disruption covered in this chapter, history is being moulded without any attempt to establish historical truths. For politicians overseeing war, narratives are increasingly experiential, of the moment, uncensored and volatile. In this manner, the digital phenomenon fundamentally complicates the process of establishing consensus about events, outcomes, and their eventual resolution. A further evolving (and uncomfortable) norm for those in power is that it is now the technologists who rule this ecosystem of process of record, mediation, distribution and content control. Indeed, much of the discussion around norm development in subsequent chapters needs to be framed by this relationship.

It therefore remains the party that controls information that can much better leverage the battlespace. Norms, however, must factor for the phenomenon's verso, the sophistication of disinformation campaigns, adversarial spoofing, deception, data denial and the evolving number of ways that these factors combine to influence battlecraft. Strategies, however, that harness influence are not new and can be as simple as scurrilous reportage such that truth is no longer an adequate defence. An enduring norm is then that actors must be ruthless in suppressing countervailing information and, in the case of the West, to do this in the understanding that democracies are usually poor competitors in this space when compared to their illiberal peers. Whispergate malware (and its perpetrators' persistent denial of culpability) provides a case in point with Russia's widespread penetration of Ukrainian systems immediately prior to their invasion's launch.[23]

---

[22] Steven Feldstein, 'Disentangling the Digital Battlefield: How the Internet Has Changed War', *War on the Rocks*', 7 December 2022, https://warontherocks.com/2022/12/disentangling-the-digital-battlefield-how-the-internet-has-changed-war.

[23] James Pearson and Christopher Bing, 'The cyber war between Ukraine and Russia: an overview', *Reuters*, 10 May 22, https://www.reuters.com/world/europe/factbox-the-cyber-war-between-ukraine-russia-2022-05-10/.

3

How Will Militaries Fight?

Realities, Ethics and Other Empirics

Russia's invasion of Ukraine has clearly upended recently held Western notions on warfare's norms. The thrust, after all, of this primer's initial evidence (taken, remember, in the year *before* that conflict's start) foresaw a more fluid and largely non-conventional set of measures that was expected to comprise the new battlefield. It focused much more upon 'grey zone' and hybrid warfare, asymmetric and below-threshold activities and generally excluded resort to conventional, overt force as the means to achieve aims. Interviews barely noted 'conventional war redux' carried out by armoured formations, flattening artillery and undertaken in a combined arms environment, all of them traits that in retrospect had already been losing stature since the end of the Cold War.

Conventional War Redux

Post February 2022, however, this matrix requires significant rethinking and, while persistent competition and its hybrid toolkit certainly remain components in current means, the prosecution of war aims through a strong conventional force is again *the* key shaping force when considering norms of warfare over the coming two decades. An enduring norm has therefore been reinforced by passing events. An issue therefore becomes the *degree* to which such factors alter the warfighting equation on the ground, the empirics of deployment, the allocation of resources, adherence to anticipated behaviours as well, of course, as the specific means and thresholds now appropriate to each contest.

The purpose of this section is to consider *how* militaries will fight in the period to 2040 and the impact of these practices on passing norms. The prevailing narrative, after all, is still based on 'old' wars (centralised, nationalistic and often ideological affairs) morphing over time into 'new' wars (decentralised episodes involving new combinations of actors and which more encapsulate identity politics). But this picture is of course too simple. It poses more questions than it answers, and this chapter therefore looks to assess the components that may comprise future battlecraft, considering the forms and means that will likely define the near term 'character' of war but do this in conjunction with conflict's otherwise softer elements (deterrence, engagement, legitimacy, will, planning, ethics, real politik) in order to chart how militaries will fight over the next two decades. The enduring norm here is that this is not all about technology. As noted by General Sir Patrick Sanders, head of the British Army, in this acerbic commentary in 2022: 'You can't cyber your way across a river.'[1]

Devoting a discrete chapter to how militaries might fight their next wars may appear ambitious but, in tackling that nexus between war's methods and its terrible nature, the authors hope that the chapter provides useful service to the whole. The analysis, after all, must deal with a fast-developing palette of issues. Interweaved into a debate on munitions and tactics, for instance, must now be the safety of communities and, notwithstanding civilians' current plight in Ukraine, the narrative of protecting peoples rather than defeating enemies.

This then is to reflect new battlefield norms and today's empirics of battlecraft. But actors' next resort to arms will certainly surprise commentators (just as Russia's actions have shocked its neighbour) and will equally likely be very different to recent episodes of how war has been waged. In this vein, a conclusion for this primer is that its initial evidence (taken in pre-Ukraine 2021 and curiously monotonal on war's new non-conventional basis) will not, of course, prove as skewed, tribal or lacking balance as it might appear from Russia's subsequent playbook of scorched earth and brazen conventional force. Ukraine, after all, represents but one war. The next conflict is just as likely to entail, for instance, a shift

---

[1]   Economist editorial, 'Ypres with AI', *Economist Special Report on Warfare After Ukraine*, 8 July 2023.

from conventional state-on-state armed conflict to much sneakier actions undertaken by unaffiliated non-state parties.

Nevertheless, settling norms today must take into account first-order observations from Ukraine and, while it may be unwise to derive firm lessons from that conflict, it is still useful to consider the exercise. First, combat operations will rarely be short-lived affairs. This is a key hypothesis, not least given its juxtaposition with America's long-held assumption that high-tech capabilities and professional personnel enable short, decisive campaigns with few casualties. Second, Ukraine's experience suggests a grinding war of attrition taking place primarily over territory that both adversaries desire. The enduring norm here remains the longer that conflict persists, the more entrenched and committed become each side's positions regardless of the means and forms of how war is undertaken. Third, wars remain difficult to end and this is only likely to exacerbate over the period under consideration in this primer.

Each of these factors individually impacts battle's broad management, from resilience to resupply, from casualties to force preservation and the reconstitution of units. While it is generally considered difficult to calculate the effects of hybrid means, it is perhaps easier to make a stab at calculating the costs of conventional war: the sum of means plus establishment plus damage plus expenditures to secure advantage approximates parties' costs. As all adversaries have found in Ukraine, long engagements require access to, and expensive replacement of, materiel as well as seamless resupply of trained personnel. Indeed, the experience of Ukraine brings into sharp focus how loss of hardware and parties' insatiable requirement for ammunition must be managed in future high intensity yet protracted, drawn-out warfare.

Norm affirmation therefore suggests a renewed significance of resilience in what is likely to be future episodes of attrition warfare, regardless of *how* that attrition and erosion are wrought by parties upon each other. This also points to a second discontinuity. No longer will 'winning the first battle decisively' be key to winning the war. As noted by Johnson, an 'inability to keep the units in the fight after significant attrition [are] the Achilles' heel of the all-volunteer professional… military'.[2]

In considering the fight, an issue for norms arises from what is the verso of the West's post-Cold War dividend. During this time, its adversaries

---

[2]   David Johnson, 'A modern-day Frederick the Great? The end of short, sharp wars', *War on the Rocks*, 5 July 2022, https://warontherocks.com/2022/07/a-modern-day-frederick-the-great-the-end-of-short-sharp-wars.

have benefitted from a three decades' learning phase during which their own military budgets have been set to target Western vulnerabilities. While Western forces were mired in counter-insurgency operations in the likes of Iraq and Afghanistan, this period of experimentation and education has allowed the West's likely adversaries to reflect on practices, observe faultlines and modernise their capabilities. Recognising that this has been to the West's detriment, its governments are resuming a periodic phase of rearmament and, on this occasion, fuelled by promised disruptions in artificial intelligence, smart machines and remote engagement that have created a revolution in expectation around war's means and shape. It is, of course, an enduring behaviour that years of underinvestment in military capability is followed by kneejerk focus on procurement, the modernising of warfare's forms and, usually, to willing embrace of technical advances.

While the West is early into this latest period of retooling and the changes in practice that it will bring, the list of required deliverables is extensive. From machine learning, the broad adoption of stealth and passive deception to the fielding of more complex and 'exquisite' fighting platforms in order to regain technical initiative over likely adversaries, the enduring norm remains that all of these means must first be deployed and then integrated into existing arsenals before each capability can be realised. This requires the lockstep development of new doctrine both for procedures and means but also for the procurement assumptions that underwrite their deployment, from deterrence and its relationship with force projection to new weapon systems, new messaging and innovative combination of methods. This then is the context for this chapter that will characterise the primer's period of interest.

Nevertheless, parties' dissembling and efforts that mask intentions will continue to create opportunity and to do this regardless of advances in how parties observe, monitor and then assess their adversaries. President Putin's textbook use of deniability ahead of Russia's crossing into Ukraine is a case in point whereby subterfuge and disinformation, long a norm and strategy of war, was still sufficient to sow doubt and obfuscate intentions. It should therefore not be surprising that neither the release of US and UK's intelligence (nor those parties' counter-messaging) was sufficient to prevent other NATO leaders from accepting Russia's narratives. Indeed, from a deterrence perspective, this is concerning precisely because it suggests that nuances in parties' messaging are more (and not less) likely to be lost in this new normal and, given Russia's own vulnerability to deception, the enduring norm remains that messages must be clearly telegraphed. Indeed, the adjunct challenge becomes that parties must take efforts to prevent their adversaries from deceiving themselves.[3]

This same information revolution and its far-reaching societal consequences mean that traditional components which have long been available for parties to signal intention (and so construct deterrence) are increasingly fragile. And while it may seem counter-intuitive in this era of being able to see and hear everything, the evolving norm must be that signalling intention and establishing deterrence is actually *more* difficult to achieve. The same difficult abstraction between war and non-war which complicates messaging also hamstrings parties' ability to make appropriate political-military responses. The verso, however, is that deterrence in a period of information revolution should really be no different from eras of previous disruptive challenge. It only works, after all, by changing an adversary's strategic calculus, each party's goal being to establish minimum deterrence (accounting for the deterring party's width, depth and sustainability of deployment) that is credible to a particular environment and the particular sets of circumstance arising. While the ingredients of deterrence (comprehension, capability, credibility, communication and, increasingly, strategies of competition) have not really changed, the issue for norms is the degree to which the current era's technology and societal developments (be they spiralling costs, the acceleration in non-lethal means of coercion or the cohesion of alliances) impacts the equation of deterrence going forward.

Informational Manoeuvre?

The book's primary evidence was almost universal in its focus on the notion of 'information manoeuvre' and the importance to norms of this change. The phenomenon was clearly a matter du jour and can be as wide ranging as simply using datapoints in all their forms to understand the operating environment better that one's adversary and thence to leverage that advantage. The aim is to shape perception as well match adversaries' options for pursuing liminal strategies, the putting in play of positions at or very near a threshold just short of open warfare in order to capitalise upon political, legal and psychological ambiguity. While the toolbox here

---

[3]   Mykhaylo Zabrodskyi and others, 'Preliminary Lessons in Conventional Warfighting from Russia's invasion of Ukraine: February-July 2022', Royal United Services Institute, December 2023, 50.

is not new, it is the available *means* to do this and a general lessening in the credibility of available information which together point to a discontinuity. And it is here where developments in connectivity and its fashioning of outputs that have given these measures immeasurably more potency. Unsurprisingly, however, the phenomenon is not clear cut. Strategies can be advanced (and blunted) by innovative or quite simple means (filters, firewalls, programmes that educate, saturate or habituate target audiences, the use of influencers, of corruption and coercion). Notwithstanding Russia's long-held reputation in this space, it was Ukraine's early performance in information manoeuvre and its tactics to control the war's narrative that initially proved a key initial contributor in seeking to offset its adversary's numerical superiority.

Information and its management in order to produce actionable context and awareness are also a factor in judging the *degree* of norm change around narratives, the more so given the very broad portfolio of sources of material and its dissemination that is now in play. It is therefore useful to rehearse again these sources (media, print, radio and television, online resources and blogs, citizen media and discussion groups) as they give colour to the empirics on the ground of the trend. Once layered and triangulated with additional information lifted from publicly available government data, from reports, hearings and conferences, the resulting mix becomes a rich resource that can be manipulated in order to shape the combat zone and its activities to the benefit of that controlling party. This represents a meaningful broadening in available supply to recently prior practice.[4] Ukraine's melding of open-source intelligence with social media, its use of command briefings and lightning-fast sharing of curated content usefully evidence these means for disseminating information and how it is already upending established practices. This has two further angles. On the strategic level, it provides adversaries with new abilities to destabilise practices that parties had previously thought secure and understood. A second consequence for conventions is evidenced by Russia's multiplication of restrictions on public freedom of expression in an attempt to keep primacy for the government's version of events.

A further norm from these developments is that parties, either near-peer or non-peer, are better able to sow ambiguity to muddy adversary's thresholds of action and to do without obvious provenance and attribution.

---

[4] Nick Reynolds, 'Performing Information Manoeuvre Through Persistent Engagement', *RUSI Occasional Paper*, June 2020, https://static.rusi.org/20200611_reynolds_final_web.pdf.

While this too is not novel, parties can now distribute tailored, segmented messaging to magnify their version of events and, in so doing, to influence narratives to their advantage. Investing in 'hearts-and-minds' and the control of events' chronology, parties seek to master political objectives in a manner that is more universal and quicker than its adversary. While all of these practices may have long precedent, the new norm is that their potency is amplified by the low-cost and innovative nature of these means, forcing planners to prepare rejoinders, to counter distortion and fabrications, to reply in kind while all of the time factoring for the threat of escalation.

Deployment, moreover, of misinformation and disinformation was once the preserve of state authorities and governments. This is no longer the case. Today, counter-factuals take seconds to manufacture and little longer to be amplified through a portfolio of feeds, bots and channels such that information is disseminated in ways well beyond the control of the original sender. The norm here is as much defined by these second order effects (and the general creation of uncertainty) as actors' deliberate actions either to contend with or exploit situations. Russia's meddling in Britain's Brexit and America's election processes are cases in point.[5] The speed and convolution of these activities and the systemic extent of their effects are the difference with previously piecemeal means.

Advances in information management suggest changes to two further norms in war's prosecution. Just as developments in societal practices may provide adversaries' unexpected opportunity for mischief, parties can less and less rely on drag from traditional practices, be that from slow-moving doctrine or from the inertia and canon that lie in establishment practices. Bodies governed by institutional thinking are innately conservative and slower to reconfigure, often regardless of the catalysts that they face. Narratives, moreover, now develop quicker and quicker. Second, norms must similarly factor for expansion in *non-lethal* means for parties to undertake conflict which are now more likely to be sinuous and low cost and which afford parties substantial opportunity to surprise. Liminal tactics (practices generally aimed at exploiting ambiguity while avoiding detection and remaining, as far as possible, clandestine),

---

[5] Amy McKinnon, 'Four key takeaways from the British report on Russian interference', *Foreign Policy*, 21 July 2020, https://foreignpolicy.com/2020/07/21/britain-report-russian-interference-brexit/ and Young Mie Kim, 'New evidence shows how Russia's election interference has got more brazen', *Brennan Center Report*, 5 March 2020, https://www.brennancenter.org/our-work/analysis-opinion/new-evidence-shows-how-russias-election-interference-has-gotten-more.

moreover, are particularly well suited for weaker and non-peer actors that are now able to impose unacceptable cost to better equipped adversaries from this expanding toolbox of means. The developing norm here is that parties' leverage is increasingly out of proportion to the delivering party's size and capacity. Ukraine's use of asymmetric means against signature assets of its adversary (whether its flagship in the Black Sea or against high-profile Russian infrastructure far away from the front line) evidences this unbalanced availability of leverage notwithstanding its position as the weaker party in its war.[6]

Liminal, unexpected activities (those courses of action perhaps previously coined 'unfair') can also exert pressure that is often difficult to counter (an example here might be the sowing of false insinuations concerning a party's leadership or legality of its operations). Liminal effects, moreover, are cumulative and long dated.[7] This too has ramifications for their use and suitability, in particular for Western democracies given their own ever-tightening decision paths where 'proof' of an adversary's involvement is increasingly a pre-condition for overt action against that party. A long evolving norm is that non-attributable liminal activities can readily freeze a party's decision processes or delay them such that options are wasted to the disadvantage of the suffering party, a trait that appears particularly acute in Western democracies.

A separate but adjacent trend is that actors, both state and non-state, increasingly seek to exploit the international community's existing rules, its red lines and thresholds as another low-cost means to create diversion and delay. International bodies and guidelines may be subverted for nationalist ends. Notwithstanding that the intended purpose of these structures is exactly to constrain parties' self-interested action, UN bodies, trade standards and international courts are increasingly organs of interest for parties seeking to subvert procedures, to obfuscate and ultimately to delay others' intervention.

---

[6] Johnathon Steele, 'Understanding Putin's Narrative About Ukraine Is the Master Key to This Crisis', *Guardian*, 23 February 2022, https://www.theguardian.com/commentisfree/2022/feb/23/putin-narrative-ukraine-master-key-crisis-nato-expansionism-frozen-conflict.

[7] Small Wars, 'Liminal and Conceptual Envelopment: Warfare in the Age of Dragons', *Small Wars Journal* (interview with Dr David Kilcullen), 26 May 2020. https://smallwarsjournal.com/jrnl/art/liminal-and-conceptual-envelopment-warfare-age-dragons.

Issues of Legitimacy and Engagement

Russia's invasion of Ukraine also makes clear that wars of conquest are no artefact of the past.[8] In considering issues of legitimacy and engagement, the assault on Ukraine has reignited debate around national resilience (for instance, its Territorial Defence Force[9]) and the training of civilian volunteers to defend homes and communities. An organised and equipped home guard capable of blunting a prolonged insurgency makes the foreign invader's goal of political control much harder. While not a new norm, Ukraine's precedent at least suggests a likely change in its priority for parties over the timeline of this primer. Resilience strategies, after all, enhance deterrence by building and then signalling national resolve. Indeed, the matter of national determination and popular fixity of purpose remains a key norm, its flip side being the relatively poor morale of Russian soldiers sent to seize and occupy their neighbour. Major incursions, after all, are rarely spontaneous. The training and marshalling of available assets remain fundamental to defenders and attackers alike, the passport to collective action as well, of course, to the management of populaces' raw fear in times of stress and attack.

An oft-cited divergence between Western democracies and autocratic parties is around actors' use of third parties to carry out their actions. As an unhelpful generalisation, autocrats continue to broaden their dependence upon proxies, surrogates, mercenaries and other unaffiliated groupings in pursuit of their aims.[10] It is difficult to frame this as a new norm. While Western democracies may appear more transparent and their leaders influenced by electorates, this is empirically not necessarily the case where, in real politik, all parties across divides have long engaged with questionable partners and in questionable practices should strategy so require. But this trait may also be in flux, the more so given the influence that charities, statutory bodies and think tanks can now have in these matters. Arguments, for instance,

---

[8]   Admiral Lee His-Min and Michael Hunzeker, 'The View of Ukraine from Taiwan: Get Real About Territorial Defense', *War on the Rocks*, 15 March 2022, https://warontherocks.com/2022/03/the-view-of-ukraine-from-taiwan-get-real-about-territorial-defense.

[9]   Mykola Bielieskov, 'Ukraine's Territorial Defence Force: The War So Far and Future Prospects', *RUSI Publications*, 11 May 2013, Ukraine's Territorial Defence Forces: The War So Far and Future Prospects | Royal United Services Institute (rusi.org).

[10]   Zoran Ivanov, 'Changing the Character of Proxy Warfare and its Consequences for Geopolitical Relationships', *Security and Defence*, 4, 2020 volume 31, https://securityanddefence.pl/Changing-the-character-of-proxy-warfare-and-its-consequences-for-geopolitical-relationships,130902,0,2.html.

in the use of explosives in built-up areas provide a case in point given the nascent consensus (for instance, the Explosive Weapons in Populated Areas initiative) that is building on that problem's mitigation.[11] Even by engaging on the subject, the US Department of Defense is indicating that Western practices in the contentious matter are more and more constrained by public pressure, a characteristic that is obviously largely absent (at least muted) in autocratic regimes.

This also feeds into the long-dated norm of legitimacy. Norms here have two rather different arcs. The autocrats' response to Western democracies being stymied by division and protest is to direct all available means to reinforce their own narrative while seeking to undermine that of adversaries. It is to pile pressure on Western electorates and other change agents (non-governmental organisations, sympathetic commentators and misinformation campaigns) in order to sow discord. Adversaries have ever more attack surfaces to create mischief and peddle influence. Intended to create legitimacy for that party's actions (the focus, for instance, by President Putin on Ukraine's 'de-Nazification'), this also adds to the shoring-up of domestic morale for the attacking party rather than any proof of buy-in or engagement from that party's constituents. Indeed, the norm here remains that each message, each campaign is ad hoc and particular to a narrative that a party is intent upon establishing at any point in time. This tends, however, not to weaken the effect of that campaign and is, after all, the long-established essence of messaging. For the UK, this might be the emphasis upon 'mission' confidence rather than 'task' confidence evident in its government's recent *Integrated Review* and, in general, its new emphasis on 'Global Britain'.[12]

Ukraine and Norm Change, 2022-2023

Notwithstanding the artificial cut-off for this primer, the start of Ukraine's counter-offensive campaign in June 2023, Russia's invasion of its neighbour continues to provide an important opportunity for parties to assess capabilities, opportunities and playbooks of the actors involved in the

---

[11] See, generally, ICRC, 'Explosive weapons: Civilians in populated areas must be protected', 26 January 2022, https://www.icrc.org/en/document/civilians-protected-against-explosive-weapons.

[12] Richard Whitman, 'UK's vision is confident, but success is a long way off', Chatham House opinion paper, 16 March 2021, https://www.chathamhouse.org/2021/03/uks-vision-confident-success-long-way.

conflict. This includes that conflict's protagonists but also, more generally, a lens into the implications of that conflict for future warfare. Moreover, it is often a war's deficiencies that are most instructive in identifying trends and lessons and, in this vein, how norms and behaviours are likely to be flexed over the term of this primer. As noted in RUSI's *Preliminary Lessons in Conventional Warfighting from Russia's Invasion of Ukraine: February-July 2022*, significant lessons will in time be drawn from the conflict but, for the purposes of this primer, it is useful even at this early stage to consider how these might influence norms.

To date, the war has predominantly been a display of old-style attrition. As highlighted by the *Economist*, it has been an 'industrial-scale contest of manpower, steel and explosives' where Russia, less than 18 months after starting its action, is thought to have suffered more than 200,000 casualties, killed and wounded, four times the number of Soviet casualties in Afghanistan, a war that lasted for a decade. This represents two and a half British armies.[13] As is often the case, commentators have been quick to compile and rank lessons learned from the conflict's first phases. The position of this primer, however, is that while the exercise may be interesting and thought-provoking, it should not detract from the uncomfortable fact that these lessons remain transient. Their impact is limited to war's character (its forms and means) and not, of course, to its immutable nature. New means are just another way of killing adversaries.

Nevertheless, three overarching developments from this conflict do bear discussion when considering norms' movement. First, there is no longer any sanctuary in modern warfare.[14] 'The enemy', notes Zabrodskyi and colleagues, 'can strike throughout operational depth. Survivability depends on dispersing ammunitions stocks, command and control, maintenance area and aircraft... Warfighting demands large initial stockpiles and significant slack capacity.' Second, the battlefield is increasingly dependent upon uncrewed aerial systems across all branches and at all echelons. Tactics and means are similarly required to *counter* such relatively new and yet already foundational assets. Third, the conflict has demonstrated the advantages of being able to deliver precision fires, the capability to engage high-value and time-critical targets and to do so with

---

[13] Economist Editorial, 'Ypres with AI', *Economist Special Report on Warfare After Ukraine*, 8 July 2023.

[14] Much of the following discussion is informed by Mykhaylo Zabrodskyi and others, 'Preliminary Lessons in Conventional Warfighting from Russia's Invasion of Ukraine: February-July 2022', Royal United Services Institute, December 2023, 2.

novel accuracy and with limited collateral damage. Precision engagement is not only much more efficient in the effects it can deliver but also allows the force to reduce its logistics tail, thereby making it more survivable. When combined with advantages in intelligence gathering, in surveillance and target acquisition as well as parties' ability to undertake effective reconnaissance (the layering of multiple sensors at even the tactical level, and the difficult for parties to conceal assets on a sustainable basis), these trends are disrupting current means but only to the extent that network-enabled warfare changed practices in the 1990s, fire-and-forget weaponry in the 1980s, and airpower developments in the 1960s.

The dichotomy here is each latest innovation usually promises deep seated, systemic changes to the battle space. The primer's recurring theme, however, is that apparently seismic developments have either a powerful verso or may not withstand the test of time. In the case of sanctuary, for instance, Russia succeeded in engaging more than three quarters of Ukraine's static defence sites within two days of crossing its borders. In the case then of uncrewed aerial systems and their importance in providing situational awareness, some 90 per cent of such systems were destroyed in the conflict's opening weeks. An evolving norm for uncrewed aerial vehicles (UAVs) might therefore be less that they are a new and revolutionary asset and rather that they are a cheap, attritable and flexible addition to existing means of carrying the fight to the enemy.

The Ukraine conflict also throws up several field lessons that, while not constituting in themselves new norms, certainly combine to catalyse new behaviours and methods. Examples here include renewed emphasis on munition stocks, on maintenance and base hardening, in cleverer communications and on a renewed focus on conflict's likely long duration. Lessons include better preparation for a protracted conflict, improved combat readiness, governments' introduction of programmes to sustain the will to fight as well as better attention paid to the 'broader fight' (the management of allies, the securing of armament supply lines et al.). Also informing norms is this embrace of uncrewed assets, the stockpiling and prepositioning critical materiel (fuel, parts, equipment and medical assets sited close to units that will need them and all held in high states of readiness).

There is, moreover, nothing like a new war to focus planners on contingencies, on training, gaming and scenario playing and the ensuring of best practice across intelligence, surveillance and reconnaissance. New priorities include the defence of logistics that is just outside of immediate theatre and, of course, the procurement of (and experimentation with) new technologies (here, for instance, loitering munitions, uncrewed vehicles in the air and in the sea, cyber and air defence networks as well as efforts to protect infrastructure assets). Taken together, the measures represent an emerging norm of dynamic and new battlefield priorities but also one that is being driven as much by past failures and suddenly visible shortcomings rather than any clever set of insights which together deliver battlefield advantage.

Even the argument for precision strikes has a verso. While their effects are obviously easy to identify, these assets will remain scarce and expensive, and their procurement constrained by complex supply issues. They too are not immune from adversarial meddling given the increasing sophistication of electronic warfare, itself a critical element of modern combined arms operations. And, while concealment may now be difficult to sustain, the enduring norm in this space remains that survivability remains viable if forces are kept sufficiently dispersed such that they then comprise an uneconomical target. The passing norm persists that forces should 'prioritise concentrating effects while only concentrating mass under favourable conditions'.[15] And other axioms have not changed. While adversaries' kill chains may still be disrupted, complicating engagement, mobility remains a critical ingredient if one's forces are to avoid attrition. Similarly several of these developments may constitute war's *improved* forms, but it remains too simplistic just to declare them collectively as a new norm given the frictions that await integration of novel systems into legacy forces, the subject of a later chapter.[16]

Empirics' Influence and Norms

War continues to be about much more than combat, the means and forms of battle and how armies fight. Indeed, the same hyperbole that can characterise latest battlefield technology may go on to influence war's broader social, political and cultural contexts and it is these, after all, that tend to determine why, where and when conflict happens. Hyperbole breeds over confidence and leads parties into risky endeavours. It is an enduring norm that this relationship determines what makes war more or

---

[15]   Mykhaylo Zabrodskyi and others, 'Preliminary Lessons in Conventional Warfighting from Russia's Invasion of Ukraine: February-July 2022', Royal United Services Institute, December 2023, 3.

[16]   See, generally, Chapter 5 (*Acquisition and Integration of Novel Systems into Legacy Force Design*).

less likely and what constitutes advantage in its practice.[17] Understanding warfare's underlying trends, therefore, and the experiences of the everyday in battlespace, helps us identify movement in its underlying norms.

Various observations merit discussion. First, while computer-driven engagement may appear to make combat more abstract (and, in that vein, perhaps more tolerable for our societies), none of this changes war's nature and the ghastliness of combat. The contradiction here is that social media and persistent news feeds then make combat paradoxically immediate and familiar. Remote lethal systems operating ever further from front lines may also make their users feel increasingly insulated from physical danger while the phenomenon obviously alters nothing for those on the receiving end of these technological asymmetries. A second empirical influence on norms relates to velocity. The speed at which machines make decisions is likely to challenge humans' ability to contribute meaningfully to processes, requiring a new relationship between the human and those machines. This certainly has the capacity to represent a new norm in warfare's prosecution. Again, however, developments have several second order consequences. This same momentum in technological innovation makes it ever harder to judge adversaries' capabilities, harder to avoid strategic miscalculation and challenging to avoid being upended by adversaries' new means, the more so as these capabilities will deliberately be hidden. Offensive cyber capabilities, for instance, rely so much on exploiting specific vulnerabilities that it becomes damaging to signal these capabilities.

Velocity has other impacts on passing norms. Just as fear and uncertainty affect volatility in behaviour, so the expectation that existing asymmetries may change very quickly impacts parties' norms around readiness, posture and response thresholds. In particular, it is movements in warfare's newer horizons (artificial intelligence, machine learning, the domains of space, deep sea and cyber) that may offer parties unexpected and disruptive advantage. It is also here that those same second order effects complicate attribution, destabilising previous constants around risk-taking and conduct. This pattern is not new. Parties developing a strategically significant lead in a particular technology have long been tempted to use that capability before being caught up by a rival. Operating at ever greater speeds, moreover, technology and new means have long put actors into

---

[17] Philip Shelter-Jones, 'Ten Trends for the Future of Warfare', *World Economic Forum*, 3 November 2016, https://www.academia.edu/29649157/10_trends_for_the_future_of_warfare.

constant states of high alert. Again, however, there is a verso here for norms whereby periods of uncertainty then tend to see parties invest in resilience, itself a stabilising trait.

Norms and conventions can also be impacted by the *perception* of advantage that movements in capabilities can induce. A case in point might be parties' seeming commitment to second-strike capabilities and its deterrence effect upon adversaries contemplating first-strike attack. An enduring norm is that military innovation is always destabilising. The notion of sophisticated drone swarms being able to attrite that same second-strike retaliation is threatening precisely because it makes first use of force more attractive if reprisal can reliably be blunted. This, however, is again to ignore the many sources of friction that exist between specifying and then deploying new means. First, long-range aerial drones are already able to navigate freely across oceans without having catalysed immediate changes to parties' force posture. Second, coordinating developments in delivery mechanisms makes it difficult for an aggressor to be sure that deterrence has been compromised. Can that party really be sure that the target cannot inflict subsequent retribution? Moreover, the portfolio of strategic means available to parties to inflict second-strike damage is always expanding (it has long been unnecessary to fly bombers around the world to undertake that first strike).

Instead, therefore, parties should generally be *less* certain how strategic benefits from warfare's new forms will be distributed. Do novel means, for instance, afford advantage to the defender or the attacker? Do they deter or incentivise escalation? In the same vein, it is also the broadening *cast* of players that complicates how norms will develop. Technology's cutting edge, after all, is no longer the preserve of developed states but has been democratised across geographies, verticals and classes of party. Seventy or more nations now operate earth-orbiting satellites. More than one hundred countries are experimenting with autonomous military technologies. The new norm is that the committed enthusiast can undertake genetic engineering in their basement.[18] Conventions must now factor for an accelerating portfolio of dual-purpose technologies across all manner of capabilities (and not just limited to the encryption, surveillance and machine learning that fills up pages of the world's press). In considering how the benefits of new means accrue, parties must calibrate their procurement plans with the additional risks of proliferation

---

[18] Ibid, 8.

and unforeseen expansion of means that dual purpose technology can now entail. Apocryphal stories of Russia repurposing washing machine processors that have been imported from neighbouring states and which then find themselves in weapon guidance systems point to the unpredictable consequences of multipurpose technology upon peer and near-peer relationships. The promise, moreover, of asymmetric advantage is becoming a more significant swing factor, the trend empowering an ever broader cohort of parties and enlarging the options that parties may use to compete. This also helps explain the complexity for war planners to navigate these developments whereby warfare's very agency is no longer the sole preserve of governments or their militaries.

Notwithstanding that much is made about battlecraft's increasing speed, campaigns are rarely decided by their first engagement and politicians must still factor for their forces fighting *protracted* wars. This is not a new norm for war planners. Populations must still be carried, national will fashioned and narratives choreographed to match much extended timelines. But this too is an area undergoing change. As Western efforts to arm Ukraine have evidenced, arms supply is both a physical and a political matter. In particular, the sophistication of modern weaponry means that stockpiles of both platforms and their munitions have all but disappeared.[19] The perils of just-in-time procurement have been laid bare. Delivery lead times for the most basic materiel will continue to frustrate planners, the more so given long-held assumptions about their forces fighting far away from their own doorsteps but, in so doing, transforming the availability, delivery and quantity of warfighting assets into an issue of renewed national importance.

While an enduring norm is obviously that planners remain forever squeezed between tight budgets, over-promising politicians and under-delivering manufacturers, procurement pinch points remain an important pivot that influence near-term norms whether through planners' periodic refocusing upon resilience or, in the case of the UK, a long period where plans hardly factored for there being existential threat to its shorelines.[20] Today, planners' talk concerns the country's logistical capabilities, its levels

---

[19] Iris Raith, 'When Arms Disappear. Europe's Persisting Problem of Disappearing Military Stockpiles', *Sphaera Magazine*, 17 October 2021, https://sphaeramag.com/when-arms-disappear-europes-persisting-problem-of-disappearing-military-stockpiles/.
[20] Joe Devanny and John Gearson (eds), 'The Integrated Review in Context: Defence and Security in Focus', School for Security Studies, King's College London, October 2021, https://www.kcl.ac.uk/warstudies/assets/ir-in-context-defence-and-security-in-focus.pdf.

of readiness, its system slack and the degree of available buffer and, given the ever-shrinking deployment timelines recently signalled for UK forces, training and human resources, including reserves. And while this list may capture the matter's current rallying points, quick-changing priorities make the substitution of issues very likely and prevent their automatic transformation into norm change.

Nevertheless, it is these moving parts that find themselves centre stage when considering developing norms and, unsurprisingly, actually evidence widespread norm *reversion*. These are the issues of every age, perhaps adjusted to reflect modern practices but systemically as familiar to the British Expeditionary Force planner more than a century ago as to the current administrator who must similarly factor for the newly rekindled importance of, for instance, second echelons and reserves. And it is here, moreover, where leaders find that long-dated economic norms, the difficult empirics of a country's balance sheet and funding priorities, have never changed. Indeed, budgetary pressures are likely only to increase as a shaping factor in force design, the more so given parties' increasing embrace of expensive technology and complex platform weaponry. The cost of the UK's two Queen Elizabeth-class aircraft carriers was reported in 2019 to be more than £7.6 billion (each with an annual operating cost just shy of £100 million).[21] The enduring norm here is that expensive munitions must make for fewer assets, less optionality and those ever-smaller stockpiles that have complicated planners' books whole decades before Russia's actions in Ukraine factored in their arrangements.

The issue of duration also impacts force design and, in complicating parties' sourcing and resupply of materiel, has planning ramifications that must impact passing norms. Just as lengthening lead times influences supply chains, it also increases the risk of interference in those processes by the fallout that arises from time to time in shifting politics and the constancy that parties can demonstrate on their domestic, international and other coalition-driven stages. This is exacerbated by weaponry's increasingly cross-border nature. The overarching norm remains, after all, that war is always a drain on parties' blood and treasure[22] and Russia's Ukrainian

---

[21] George Allison, 'How Much Does a Queen Elizabeth Carrier Cost per Year?', *UK Defence Journal*, 18 July 2021, https://ukdefencejournal.org.uk/how-much-does-a-queen-elizabeth-class-carrier-cost-per-year/.

[22] Adam Staten, 'Russia Spending an Estimated $900 Million a Day on Ukraine War', *Newsweek*, 6 May 2022, https://www.newsweek.com/russia-spending-estimated-900-million-day-ukraine-war-1704383.

invasion has unsurprisingly reiterated the very substantial investment in stocks, resupply and planning that is required for conflict.[23] Duration, of course, has particular norm-affirming effects on those participants over whose land that war is fought.[24] It is thus not at all surprising that Ukraine has embodied the key characteristics of resilience, engagement and morale.

If a verso exists to this observation, then it is that planners must match their war's extent and intensity with available capacities and resources. Indeed, an enduring norm remains that the pace and depth of states' participations must align with the available (and finite) resources at hand for that war's prosecution. While this is not new, and while changes to war's landscape mean disruption of all parties' decision sets, these changes do not necessarily happen in lockstep, exacerbating difficult procurement choices, epecially given recent battlecraft that is based upon multiple episodes of quick and active fighting.

Understanding norms is also complicated by new means of harnessing *non-traditional* capabilities to augment parties' ability to fight. These may include internal affairs forces, law enforcement assets and other local paramilitary formations that allow parties to operate outside their traditional military footprint. Russia's use of private military companies (directly on its own balance sheet but also organised by state-owned commercial bodies) is a case in point and, notwithstanding that mercenaries have long been a feature of the battlefield, their use suggests at least a reinforcing norm, complicating the shape of deployable forces, their funding and delineation. Russia's Wagner Group experience, moreover, highlights the difficult balance that has long characterised these arrangements, their degree of independence, integration and the control of third parties in a state's concentrated efforts to win the fight.

Russia's use of mercenary assets also highlights an adjunct norm around parties' plans to conduct sustained operations *outside* their borders with just a small, scalable footprint positioned on the ground. Russia's poor performance in the opening phases of its invasion of Ukraine suggests that this assumption rarely holds. It also refreshes how hubris, over confidence and the misreading of signals can undo the most detailed planning.

---

[23] David Maccar, '2 Million Rounds Headed to Ukraine from American Ammunition Manufacturers', *Free Range American*, 7 March 2022, https://www.newsweek.com/russia-spending-estimated-900-million-day-ukraine-war-1704383.

[24] Madeline Halpert, 'Russia's Invasion Has Cost Ukraine Up to $600 Billion, Study Suggests', *Forbes*, 4 May 2022, https://www.forbes.com/sites/madelinehalpert/2022/05/04/russias-invasion-has-cost-ukraine-up-to-600-billion-study-suggests.

Indeed, successful battlecraft eventually relies upon the same prerequisites that underpin all regular forces, the primacy of assured logistics, local engagement and the buy-in of those tasked to deliver politicians' aims.

Understanding battlefield empirics also requires consideration of parties' casualty management. Several drivers shape its processes. First, it may be that the notion of casualty aversion exists more in the minds of militaries and politicians. Repeated polls suggest the issue remains a low priority for the public. The passing norm should therefore be that this casualty calculus is more nuanced than suggested by media coverage. Efficient casualty management is nevertheless a key component to the unwritten contract upon which personnel serve. It informs morale and provides a benchmark to how parties treat and value those serving in its forces. Casualty 'aversion' then catalyses issues around training, replicability and the ready replaceability of service personnel. It concerns the imperative of mass and the maintenance of institutional knowledge. As an issue, it also finds itself entangled with hardware considerations (the dynamic equation of losing personnel and losing kit) and issues around the availability and replenishment of war materiel (can personnel be adequately protected in their mission and is this mission suitable?). Given the primer's timeline, conventions around preserving life and avoiding casualties will also be tied into the adoption of more remote warfare where the introduction of autonomy incrementally replaces humans by machines, especially in dangerous, difficult and repetitive tasks (and thereby reducing the chance of losing soldiers).[25]

Liminality and Norms

The principal question for this chapter is how militaries might fight, 2025 to 2040. In this vein, the evidence taken for RUSI's initial research project repeatedly homed in upon developments in parties' *non-lethal* playbook and the wide range of available non-kinetic means that this might entail. Commentary here noted the growing role of private military companies, of state-sponsored extortion, financial and currency warfare. In considering methods, it also highlighted adversarial tactics around the 'fait accompli' whereby parties' actions are designed to exact small gains from an opponent using sufficient credible force to compel the opponent to accept that loss rather than retaliate in kind.

---

[25] See Chapter 6 (*Autonomy and Thresholds of Supervision in Lethal Engagement*).

Notwithstanding Russia's overtly conventional tactics used in Ukraine, commentators pointed again and again to the last decade and its trend towards liminality, the amplification of actors' practices through diversionary tactics and temporal ambiguity. In particular, China and Russia have been successful in their tactic of 'reflexive control' whereby targets are encouraged to act in the interests of an actor without realising that they have done so.[26] But the verso here is starkly provided by Ukraine's experience. Pivoting force structures away from conventional to liminal means contradicts long-held battlefield fundamentals and, in light of Russian tactics, now appears less responsible. The enduring norm is that militaries still require the equipment and mindset to win their wars, to attrite adversaries in order to win the fight. Indeed, the primacy of this norm persists *regardless* that wars rarely end to order.[27] It even reinforces the norm around leveraging an 'adaptation gap', the degree to which parties must adapt in the face of technology developments across all assets that might have military use. This is now considered below.

Norms around Adaption and Innovation

This all presages an interesting debate around how conventions should factor whether adaption will be more relevant going forward than innovation, the crux being how parties can further leverage their existing arsenals by modifying their use and combination. Rather than encouraging forces endlessly to innovate, it might be that the norm should instead emphasise the premium to forces' *adaptability*. Quick decision-making and in-the-moment adaption are important advantages in battlecraft. Adaption of course is not new and provides the spine for much of Western forces' training. Decisions undertaken in moments of stress and where events are moving very fast are a case in point where commanders stray from expected norms in exploiting whatever capabilities are available.

Adaptive examples include refurbishing and repurposing existing hardware but it also might extend to media manipulation, non-traditional exploiting of actors' errors as well as the repurposing of everyday consumer

---

[26] Peter Mattis, 'Contrasting China and Russia's Influence Operations', *War on the Rocks*, 16 January 2018, https://warontherocks.com/2018/01/contrasting-chinas-russias-influence-operations/.

[27] Margarita Konaev and Polina Beliakara, 'Can Ukraine's Military Keep Winning?', *Foreign Affairs*, 9 May 2022, https://www.foreignaffairs.com/articles/ukraine/2022-05-09/can-ukraines-military-keep-winning.

systems in clever efforts to combat the adversary. Conflict, after all, throws up learnable lessons that are applicable to the short and medium terms alike and, while these should not prompt knee-jerk change in norms, they are *cumulatively* important in refining behaviours, in provoking revision as well as resetting degrees of tolerance in how parties plan their operations. This might include the command of troops, the safeguarding of soldiers against the adversary's own innovation and, germane to Russian difficulties early in its campaign, the consequences of insecure communications and other instances of just poor practice. Adaption's timeline, moreover, can be very quick. Unlike the byzantine processes of procurement, adaption may involve simple overnight tweaks, the embrace of quite straightforward adjustments to blunt, for instance, a particular advance observed in an adversary's capabilities.

This is not to sideline the importance of innovation and its potential to influence norms. Adaptive practices find solutions using already established systems and assets whereas those who are innovating will go beyond current practices to find new and untested workarounds to battlefield issues. Innovation perhaps works to longer timelines than adaption but can have as transforming effects. The developing norm here is that innovation usually requires already established and process-heavy ecosystems where leadership exists to push its deployment down to the front line. To be successful, it requires long-term and persistent support. Empirically, it needs patience and leeway. Its verso revolves around the *degree* of challenge that innovation involves in its duration (here, the length of time between concept and deployment), its sustainable funding and institutional support, the extent of required interoperability and, as always, political considerations and lines of accountability in the delivery of these programmes.

Occupation and Norms

In considering how militaries will fight, Russia's invasion of Ukraine reveals the divide between what expected behaviour and that which norms suggest will occur. Nevertheless, while shocks and surprises happen, norms are eventually shaped by war's passing forms and events. An example is useful. The conventional playbook of artillery attrition undertaken by Russia has been quite different from the hybrid toolkit imagined by thought leaders in advance of Russia's invasion. It demonstrates, if demonstration was necessary, that onslaught and thuggery remain the key constituents of warfare. And while norms going forward must reflect this basis, conventions around law adherence, ethics and morality also remain relevant, even if they set actors apart in how war is prosecuted. Indeed, divergence from previously normalised behaviour is itself a long-dated norm, the more so in times of conflict and political stress, such deviation being reflected in the moment's empirics of fighting and how parties act in their efforts to win. It further raises the question, however, around the degree to which norms are more a Western construct and how they really factor in the decision-making and actions of autocratic parties.

With regards to norms around occupation, Russia's targeting priorities in Ukraine have certainly tested assumed norms. Examples abound, whether in its use of illegal munitions in Bucha, Hostomel and Borodyanka, the extensive targeting of civilians or its destruction of civilian infrastructure.[28] Russia's activities are certainly inconsistent. The Kremlin's early decision to seize Ukraine's reactors, for instance, might have been an effort to control the host's electricity generation but initiated unpredictable environmental consequences that directly undermined accepted norms from which it itself benefits. While Western parties usually exhibit tacit acceptance for norm adherence, Russia's activities evidence instead the much less predictable actions of autocracies that can operate without domestic constituencies to appease.

Weapon Innovation and Norm Change

Russia's war in Ukraine has allowed those in manufacturing and procurement to assess the performance of weaponry, to consider this relative to its peers under a range of live scenarios and to do this almost in real time. The war's extension of war's forms has also provided users with an opportunity to convert the *potential* of their systems into tested and proven performance and norms must empirically reflect this quicker cycle. Innovation and adaption of means, moreover, have been accelerated across all of war's domains and practices, across defence and offence as well as in parties' ability both to absorb strikes but also to deny adversary's ability to bring corresponding force to bear.

Examples abound to illustrate the trend including deployment early in the conflict of the Javelin, a compact, shoulder-launched, portable

---

[28] Lorenzo Tondo and Isobel Koshiw, 'Ukraine destruction: how the Guardian documented Russia's use of illegal weapons', *Guardian*, 24 May 2020, https://www.theguardian.com/world/2022/may/24/ukraine-destruction-how-the-guardian-documented-russia-use-of-weapons.

rocket launcher, the system's fire-and-forget, anti-armour capabilities initially playing havoc against traditionally deployed Russian tank assets. Employing two motors (and so preventing the operator being knocked out by back blast), its munition is a twin-action-shaped charge capable of attacking tanks' less protected upper surfaces. Similarly impactful to Ukraine's defence was the earlier deployment of the Stinger platform, also human-portable but deployed as an anti-aircraft system that is capable of locking onto the heat signature of its aerial target. Employing rudimentary autonomous tracking, the system can engage to an altitude of more than 3,000 metres. Ukraine has also fielded Switchblade drones, another expendable UAV but one with limited although important loitering capabilities and an ability to lock onto its target. In a clear signal of these technologies' new reach on the battlefield, the weapon is able to optimise performance by toggling between an on-board camera, its GPS and its function of operator-aided target tracking.

Although not in themselves individually disruptive, the trajectory of these technologies has certainly shaped how both attacker and defender must reorganise their actions and adjust priorities. Their deployment is doubly significant given the knowhow and experience base that their use creates in what is a developing field of warfare. Finally, to this point, norm adjustment is obviously to be seen on both sides of a battleline. While its deployment of heavy artillery (its Grad, Smersh and Uragon systems) conforms to long-held battlefield practices, Russia has been similarly active in deploying new and adjusted means, both in new UAV platforms but also in otherwise traditional means, its POM-3 mine with seismic sensors that are triggered by proximate footsteps.

4
How Will Conflict Be Waged
The Dynamic of Conventional and Asymmetric

An overarching norm change in battlecraft arises from the relentless democratisation of means that can be used to prosecute war. Parties, particularly those that may be near-peer to an adversary, are able to pursue their strategic aims with a toolbox of means that is developing very quickly.[1] Indeed, ways of causing mischief and an ability to degrade how an adversary functions are increasingly in reach and, while Ukraine demonstrates that kinetic means dominate norms in the specific conflict, it is the adoption of matching *non-kinetic* options that requires norms be refined and tuned in order to represent the period of this primer.

An aim of this chapter is therefore to frame how these new avenues of opportunity (economic, informational, social, reputational, commercial, political) can inflict costs upon an enemy. All of this depends, of course, on the degree of interconnectedness observable in states, their people and their militaries from the adoption of technology and the embedding of new practices. While not new, these usually reflect parties' broadening commercial links and their expansion across a growing matrix of partners (and the conflicts of interest that these relationships may then entail). They also reflect a new transparency under which these arrangements are being made.

---

[1] Reyes Cole, 'The Myths of Traditional Warfare: How Peer and Near-Peer Adversaries Plan to Fight Using Irregular Warfare', *Small Wars Journal*, 28 March 2019, https://smallwarsjournal. com/jrnl/art/myths-traditional-warfare-how-our-peer-and-near-peer-adversaries-plan-fight-using.

Norms' Velocity

The release in 2021 of the UK's *Integrated Review* coincided with an end to two decades of Western counterinsurgency operations in Iraq and Afghanistan and a shift in defence establishments' focus from paradigms of 'endless war' to, once again, more conventional models based on peer and near-peer conflict. It might be expected, therefore, that norms over the period of this primer should be dictated directly by geo-political context. This, however, would require the factoring of a meaninglessly broad list of issues that change month by month. It also assumes that these factors are capable of clear and easy abstraction. Any such list would also be fickle and volatile, driven by the regularly changing priorities of parties that are in permanent pursuit of hegemony (whether that be economic, technical, societal, military or doctrinal). Besides, these matters are generally too nebulous to translate into clear norms and tend instead to reflect each party's particular history and set of narratives.

A driver here is the ability and, more importantly, the *will* of parties to marshal assets against adversaries' vulnerabilities, to do this in a manner that creates systemic advantage and to do this regardless of physical distance.[2] How this impacts passing norms, however, becomes complicated. In the West, for instance, sources of long-dated friction include the frustration of five-year planning cycles, procurement and delivery issues arising from 'the fighting of tomorrow's war with the last war's capability' as well as planning horizons based upon the near-term at the expense of long-dated outcomes.

It is military practices, however, that remain the key determinant in catalysing changes in behaviour. Just as duration and velocity can distort planning horizons, how force is organised and then deployed (the issue, for instance, of 'jointery' arising from commands that are shared by two or more branches of the same combat force) impacts how norms may shift in response to developments in battlecraft. Changes in military processes, however, have traditionally been slow and drawn out, their behavioural impact usually dependent on where in battlespace those changes were being undertaken.[3]

---

[2]   Daniel Riggs, 'Re-thinking the Strategic Approach to Asymmetric Warfare', *Military Strategy Magazine*, Vol. 7, No. 3, Summer 2021, https://www.militarystrategymagazine.com/article/re-thinking-the-strategic-approach-to-asymmetrical-warfare/.

[3]   Benjamin Jensen and Matthew Strohmeyer, 'The Changing Character of Combined Arms', *War on the Rocks*, 23 May 2022, https://warontherocks.com/2022/05/the-changing-character-of-combined-arms.

Military practice has therefore been a laborious and inefficient change agent but is now undergoing profound revision, driven in the main by the new means (and their complexity) that comprise this primer's analysis, but also the integration of digital, often machine-based processes that are becoming the new backbone of battlefield operations. The authors are always concerned to identify arguments' versos. Here, it might be that digital infrastructure is fragile, its outcomes too readily compromised by partial, duplicatory, contradictory and erroneous data. The introduction of these new means might be particularly prone to institutional inertia, to poor general understanding of newly complex systems and their possible reach as well as insufficient skills across receiving organisations. For transformation to occur, moreover, the evolving norm must be that trying, failing, and reworking, practices that are often a challenge to militaries' culture, must become an integral component to deployment.

Planners, after all, are not alone in witnessing the destruction of Russian mechanised formations in Ukraine in an apparent shift of power to the defence, parties' improvements in lethality, the integrated use of next generation weaponry and what this mix of effects might herald. Several questions arise. In the case of armour, for instance, it is whether ground can be seized and an entrenched enemy destroyed (especially in urban terrain) without the use of that asset. Battlecraft is tricky and closing with an enemy with or without tanks has long been a difficult task fraught with exogeneity. Luck, brilliance, the unforeseen and the random all have their part in outcomes, the point being that degrees of norm change can only be assessed by considering trade-offs and taking the long view. Today's potentially disruptive innovation can turn quickly into tomorrow's old news. Battlefield success will still revolve around better logistics, better allocation and availability of materiel and, of course, the calibre of those leading and using these assets.

The Hybrid Toolkit

The planners' challenge in responding to war's *hybrid* activities comes from how best to delineate actionable components from what is a considerable portfolio of available means to disrupt adversaries. Below-threshold activities are rarely new and seldom consistent and this muddies their classification into sets of conveniently abstracted behaviours upon which the planner can build and test norms. For the purposes of this primer, grey zone actions encompass those activities that fall short of high-intensity warfighting and which are inherently ambiguous such that it is difficult to establish their potential for escalation. This is not particularly helpful. Misinformation, for instance, is today so commonplace that arguments can be made for it to be downgraded from a specific grey zone activity to routine statecraft. More generally, hybrid means should be regarded as a long-established and elastic approach to seeking advantage. Using inventive, sinuous, creative initiatives to defeat an adversary is, after all, a strategy as old as warfare itself and long established as an enduring norm. Hybrid's attraction lies in this lack of definition. Similar to autonomy in lethal engagements, it sits along a continuum and adversaries should expect to see it deployed on its own, in parallel with kinetic means or at any place along the mix of these two states.

Ukraine provides a case in point with Russia's execution of old-school kinetic assaults being undertaken very much in conjunction with hybrid means including cyber and electronic warfare, information and propaganda operations, energy blackmail and commercial extortion.[4] Empirically, however, the norm has been that sub-threshold activities were usually better played by the party that had already achieved some technical superiority (qualitative advantage) with the aim then of hybrid activities supplementing conventional military mass (that is, quantitative advantage whereby the many beats the better).[5] Two considerations arise. First, hybrid means is now likely to be materially cheaper than the deployment of expensive conventional military hardware and therefore a preferred means for non-peer, near-peer or otherwise disadvantaged parties. Second, obfuscation and deniability are ever easier to sow in today's digital age where narratives can be created ahead of actions being undertaken. The enduring norm here is that it remains hybrid's breadth, competitive cost and degrees of surprise that make its toolkit so appealing.

Adversaries can therefore deploy assets from this portfolio of techniques and do this in concert with other means. Indeed, it is this diversity of attack points that can deliver outsize consequences whether from disinformation and misinformation programmes, the use of private military companies in covert and overt operations, deployment of

---

[4]   Economist editorial, 'What Is Hybrid War? And Is Russia Waging It in Ukraine?', *Economist*, 22 February 2022, https://www.economist.com/the-economist-explains/2022/02/22/what-is-hybrid-war-and-is-russia-waging-it-in-ukraine.

[5]   Benjamin Jensen and Matthew Strohmeyer, 'The Changing Character of Combined Arms', *War on the Rocks*, 23 May 2022, https://warontherocks.com/2022/05/the-changing-character-of-combined-arms.

cyber, the support of terrorism or other irregular activity. Hybrid means include electronic attack, infrastructure meddling, deniable but also overt economic means (for instance, Russia's weaponising of its energy policy), criminal means, the hijacking of cross-border institutions and global standard setting organisations, political interference and agitation. It may also include bribery and corruption, and, of course, a combination of these measures such that no single hybrid operation is undertaken in isolation.[6]

Ukraine has demonstrated that parties will generally be agnostic in their switching between hybrid, kinetic and other non-kinetic activities and it is the availability of such *combinatory* approaches which suggests that warfighting over the coming two decades will less and less be defined as plainly conventional, hybrid or asymmetric. Given also that the Ukraine conflict is but one conflict, an adjunct norm here becomes that near-peer leverage arises precisely from that weaker party's parcelling up of warlike activities as mere state policy such that the stronger party is unable to deploy its stronger conventional hand without being the first to cross over into outright hostilities (with the societal and geo-political implications that this likely entails). This is an uncomfortable trigger point for parties to cross and, as such, the norm will be that hybrid 'wars' are just as likely to extend into 'forever wars'.

Expansion of warfighting by non-war means is not new and is therefore a long-dated norm in its own right. What is interesting, however, is then the degree of lockstep evident between means deployed and the changes in behaviours that they catalyse. Is there a way to discern which means are more impactful? And while this may together elevate the importance of sub-threshold activities, it may also cause parties to reframe the role of conventional means in their force posture, the more so in times of slim budgets and the need to tinker with priorities. This too is not new. The standing, after all, of the Fulda Gap as a uniquely strategic tripwire between the West and East has been diluted for decades.

If this is the case, it is useful to detail other activities that may comprise sub-threshold conflict. In addition to the manoeuvres listed above, the hybrid battlefield today now encompasses parties' electoral landscapes, their social media and domestic infrastructure. It is the global financial system that provides a new attack angle in non-war competition between

---

[6]   Arsalan Bilal, 'Hybrid Warfare - New Threats, Complexity and Trust as the Antidote', *NATO Review*, 30 November 2021, https://www.nato.int/docu/review/articles/2021/11/30/hybrid-warfare-new-threats-complexity-and-trust-as-the-antidote/index.html.

states, commercial entities and other actors. Indeed, all of these developing touchpoints initiate their own and quite individual set of consequences. The developing norm is clearly that hybrid competition has an evolving set of effects. The paradox for this analysis, of course, is that this acts to destabilise current conventions. Collateral can now be created in quite new and perhaps disrupting forms as traditional battlespace expands, unsettling other previously stable relationships and behaviours. It defies guesswork, messes up expectations and, in so doing, tests commanders' behaviours and the doctrine that shapes them.

Hybrid measures therefore increases scope for misunderstanding and misstep. It is not even clear that they are a stratagem and, just as attribution of attack is likely to be ever more confusing, establishing hybrid actions' primary and second order effects is likely to be similarly equivocal given its extending portfolio of means, from blockading grain shipments in the Black Sea and a misinformation campaign (one that wrongly posits breaches in international humanitarian law) to cyber sabotage of civilian infrastructure. The statement is reinforced that adversaries no longer fight in manners previously expected of them. It becomes the norm and, while this too may not be new, a lesson from hybrid warfare is that traditional battle is no longer, Ukraine notwithstanding, likely to be the decisive encounter. Instead, components to victory might now be very wide, combinatory and unexpected. Examples of asymmetric warfare are just as likely to include actions designed to displace populations, state-sponsored initiatives encouraging corruption and the programmed pursuit of intimidation in order to influence power platforms.

Economic Warfare and Norm Change

The fact that Russia appears not to have factored for the threat of sanctions in its Ukraine calculus is unsurprising as such means have long been available but usually deemed ineffective. More than 10,000 people or firms are currently subject to American sanctions involving over 50 countries and some 30 per cent of the world's gross domestic product.[7] This may be a norm but one that is neither as efficient nor in any way as targeted as suggested in the media. In the case of Ukraine, moreover, any norm change is less about individual Western measures to blunt Russia's economy but

---

[7]   Economist editorial, 'A New Age of Economic Conflict', *Economist*, 5 March 2022, https://www.economist.com/leaders/a-new-age-of-economic-conflict/21807968.

rather the unprecedented scale of these coordinated, concerted efforts. The 2022 sanctions against Ukraine's aggressor represented a whole new dimension of economic action, the effects of which may be considerable but very unlikely to be as anticipated at their outset. An enduring norm of warfare (and an effective counter to the means) remains, after all, that parties can endure considerable pain over the long timelines of a conflict and it will certainly remain the case that countries can suffer severe economic dislocation. Indeed, Iraq's oil-for-food sanctions beset its population for more than two decades without regime change or material repercussion.

While sanctions will remain an important component of deterrence and penalty, their imprecision provides useful insight into their shortcomings as a strategic tool. The *Economist*'s initial assessment of Western economic measures should reinforce this skepticism. Early assumptions around sanctions forecasted the 'instant immiseration of a large economy', at that time the world's 11th largest economy, with its inference being that an unprecedented programme of economic measures would cause potential belligerents to recalculate the costs of initiating conflict.[8] This has not played out notwithstanding, importantly, that Russia remains largely removed from previously integrated global supply chains. While the calculation here was that sanctions would have profound and abrupt consequences upon the aggressor's economic activities (especially in that country's forward ability to manufacture and replace sophisticated materiel), the world's miscellany of supply chains and the complexities of parties' real politik have meant that Russia has been able to replace component sources and shift its business to other willing partners.

Other economic behaviours are affected by Russia's experiences. In outlawing commercial activities with Russian banks and expelling them from the global-payments plumbing, Western allies' actions have likely created an unexpected new norm arising directly from actions to hobble cross-border flows of money into adversaries' economic spheres. Parties which are not outright allies of the US may now think twice before accumulating US Treasury bills out of fear these assets may get sanctioned, frozen or cancelled.[9] The more that sanctions are deployed, the more aggressing parties will presumably seek to avoid relying upon Western finance.

Second, economic warfare is likely to encourage the building of new and quite separate spheres of economic influence and, in time, an

---

8   Ibid.
9   LV Gave, 'The End of the Unipolar Era', *Gavegal Research*, 11 May 2022.

unpredictable delinking of anxious parties from Western-led financial systems. The West's Russian experiment, after all, has already prompted several second order effects. Developed economy government bonds, for instance, proved anything but safe in the first half of 2022. Similarly, seizure by a central bank of another sovereign country's assets has legal ramifications concerning property rights and the guarantees provided by the rules of law. At the very least, new safe destinations for emerging markets' excess capital will emerge. Sanctions therefore force various remedial actions upon parties which, in turn, lead to evolving changes in norms for parties on either side of each sanction-relevant transaction.

Russia's sanctions package and these programmes' second order effects also create new planning complexities that have arisen from this century's extraordinary developments in global trade and commercial interconnectivity. The evolving norm is that it is increasingly difficult to predict the efficacy of these economic packages precisely because they require the compliance of every involved party. That same interconnectivity means that there is a runaway number of intricate parties that must be factored for in the incentivising, construction and then monitoring of these schemes. The motivation to cheat is high. The opportunity to dissemble is high. The introduction, moreover, of these measures tends to be weakeningly incremental with subsequent rounds of a sanctions package not then being adopted by one or other interested parties, so affecting the efficacy of the overall programme and moving further and further outside the power and intentions of the initiating parties.

A further problem remains that sanctions' pain is usually unevenly felt, either by the target or, as importantly, in the often fragile cohort of sanctioning countries. In the case of Russia, after all, it is very likely that the state has been assisted by dissenting parties happy to evade Western sanctions. The norm remains, of course, that a party under sanction can quite quickly move to blunt those sanctions' effects by rationalising its own procurement practices, streamlining supply chains, investing in sanctions subterfuge and, wherever possible, by initiating state-wide programmes that accelerate import substitution.[10] Nevertheless, these measures involve changes in behaviour and, as such, constitute an evolving variation to a previously accepted norm. Nor is this process straightforward. Kaushal, for

---

[10] Sidharth Kaushal, 'Can Russia Continue to Fight a Long War?', *RUSI Long Read*, 22 August 2022, https://rusi.org/explore-our-research/publications/commentary/can-russia-continue-fight-long-war.

instance, notes that Russia was able to achieve effective swaps for previously imported Western goods in just seven out of more than a hundred categories of key defence equipment following the sanction programme instigated by Western states after Russia's annexation of Crimea in 2014.

The norm is further complicated, moreover, as parties' commercial strategies mitigate sanctions in line with their own priorities and real politik. Sanctions' success depends in part, after all, upon the underlying products' fungibility (the swapping and repurposing of similar products, often for a military end). Of course, not all products are fungible in the open market, and bills of material for military platforms will very often incorporate bespoke individual items that are directly impacted by a sanction and must be swapped, bartered or designed out if that product is to be deliverable. A second verso, moreover, is that an upper-middle economy such as Russia will long have embedded dual-use technologies and alternative supply chains into its economy, further muddying attribution and blunting requirements that the country immediately pivot its behaviours.

Sanctions, moreover, may be bypassed in any number of ways. An example is China's current experimentation with a digital version of its currency. The e-CNY has seen more than 260 million transactions that were worth some $12 billion in the 24 months to June 2022. While the rise of alternative platforms may or may not help settle international payments at a fraction of the cost of the current corresponding-banking model, innovation in such global systems might reduce parties' vulnerability to sanctions and, in the future case of 'another Russia', dull further the precision of America's own financial weapons. Initiatives such as China's e-CNY initiative have the scope to provide non-US aligned parties with a platform to match America's sanction programmes and, as such, represent a driver of norm change, transactions in e-CNY taking place across the balance sheet of China's central bank in a system that allows participating governments to retain oversight and control.[11]

In considering norm change, however, Western democracies (here defined as the high-income democracies) still account for more than 40 per cent of global output at purchasing power parity and nearly 60 per cent at market prices as at 2022. The corresponding figure for China remains less than 20 per cent on both measures (with, incidentally, Russia being a de

---

[11] Economist editorial, 'The Digital Yuan Offers China a Way to Dodge the Dollar, *Economist*, 5 September 2022, https://www.economist.com/finance-and-economics/2022/09/05/the-digital-yuan-offers-china-a-way-to-dodge-the-dollar?

minimis fraction of these figures). The West, moreover, continues to issue all significant reserve currencies. During 2022, China held more than $3 trillion in foreign currency reserves while the US held almost none. Instead, the US Treasury can print them. The verso remains, nevertheless, that while economic norms may appear volatile and prone to kneejerk change, the deep-seated forces of economic and geo-political inertia will persist long after the period of interest for this primer. Empirically, moreover, isolated autocracies appear over time to address the very supply and manufacture imbalances that sanctions are intended to occasion. Indeed, vested interests and geo-politics remain important and thoroughly reactionary drivers that temper wholesale adjustment to the status quo but also, by extension, to the long-dated set of advantages currently enjoyed by the West.

A quite different development (but one still with ramifications on sanction regimes) deserves review. Correlation is no longer clearcut in the relative utility of means between the expensive and exquisite (here, complex weapon systems) versus home-grown and budget. The convention remains that low-tech solutions can often be rigged up using a kaleidoscope of fixings and componentry. While too early to draw conclusions, Ukraine's early success in deploying comparatively cheap technology to overcome the costly, complex weapon platforms of its adversary might further compromise the efficacy of sanctions that target, for instance, a party's supply lines and recourse to technically advanced materials. Assuming an equal will to win the fight, a developing norm might be that that home-grown, cheaper means can exact asymmetric damage on those rich states armed with costly sophisticated platforms, the more so in protracted conflicts, and raising disproportionately the price of an aggressor's decision to commence battle.

This has consequences for parties' deterrence but also the general economics of war. Indeed, the experience of Ukraine would seem to indicate some general leveling up in battlecraft (here, the character of war) regardless of a party's relative investment in means, as well as a narrowing of the capability gap that might traditionally have been expected across warring parties' force structures. While Russia and Ukraine's deadlock in the summer of 2023 is not coincidental, the associations and behaviours that underpin this stalemate are less clearcut. While the examples of Iraq and Nagorno-Karabakh confirm the edge provided by technology in those conflicts, the experience of Ukraine has suggested a tilt back towards traditional practices, attrition and grind where technology is either not available, has been nullified or is otherwise irrelevant.

Nevertheless, the first 14 months of the Ukraine conflict have not just been a brutal artillery contest. In the economic sphere, it is already a much larger strategic confrontation, destabilising age-old trade relationships and upending long-established supply lines across the world for decades to come. Indeed, in terms of warfare's economic norms, a whole new battleground has emerged, one with which a recently globalised economy is struggling to contend.[12] A new norm is therefore that sowing economic disruption *outside* the imposition of sanctions appears to be easier and more far-reaching for those parties able to employ adversarial strategies in this space (Russia's weaponising of oil and grain exports and the subsequent volatility in these two markets). In 2021, Ukrainian foodstuffs fed some 400 million people worldwide (with the Middle East and Africa being key customers) but food shipments sent through Black Sea ports have become an overt weapon of war.[13] Nor is an adversary's opportunity set limited to energy or food-producing governments. Parties' sudden interest around undersea cables is a further case in point that has prompted similar volatility in rhetoric, markets and, over time, behaviours as parties rush to shore up possible areas of economic vulnerability, whether through new technologies, new alliances, diversified markets or better buffers in their own arrangements.

The Norms of Defence and Offence

Ukraine's defence of its homeland has triggered other reminders of battlefield empirics. First, it is refocusing attention on the long-fluid equation between attack and defence.[14] While the relative primacy of these two states may appear tied to a particular war's particular character, the inference is that specific scenarios and battlefield conditions determine the passing predominance of whichever state, offence or defence. Certain observations are still useful, notwithstanding that winning the fight remains exogenously influenced by good fortune, by one side's better tactics and leadership, equipment, morale and logistics.

---

[12] Johnathan Chang and others, 'The Economic Front in Russia's War against Ukraine', *WBUR On Point*, 8 March 2022, podcast, https://www.wbur.org/onpoint/2022/03/08/economic-war-and-russia-ukraine-conflict-sanctions.
[13] Peter Apps, 'Black Sea Grain Battle Could Define Ukraine War', *Financial Times*, 30 May 2022.
[14] TX Hammes, 'The Tactical Defence Becomes Dominant Again, *Joint Force Quarterly*, 103, 14 October 2021, https://www.960cyber.afrc.af.mil/News/Article-Display/Article/2810962/the-tactical-defense-becomes-dominant-again/.

While the relative hazard of attacking or defending may have ebbed and flowed in lockstep with moves in technology and doctrine, it has also been influenced by behavioural considerations. First, battlefield operations are always situational. They are also transactional, depending upon the interactions of people and materiel. Second, outcomes during the Ukraine war's first phase were, in retrospect, considerably driven by intangibles such as adaptability, spirit and effective collaboration in the face of its people's existential peril. Russia's invasion rekindled long-dated norms of nationalist pride, stoked by national affront, outrage and a call for revenge. Early wins fuelled Ukrainian confidence, with new equipment and new practices quickly suggesting the beginnings of norm change right across battlefield practices, particularly around the role of the resolute, determined defender and its success using new combinations of hardware to offset Russia's offence (whether because of that party's very poor planning, poor training and absent leadership or, more likely, a combination of these factors). The year 2022 established that prepared defensive systems could at least match an aggressor's purportedly more sophisticated platforms.[15] This set of circumstances has now been replicated by Russia as it digs in and prepares ground in order to contend with Ukraine's counter-offensives over the summer of 2023.

In fact, swings between offence and defence can be quite transitory, even across prepared positions. Their relative cadence more depends upon each battle's context and how warring parties prioritise the manner in which they will fight, the balance parties achieve between battlecraft's subjective characteristics (leadership, determination, resilience and national will) and the objective matters (tank numbers, missile availability and efficacy, battle plans and logistics) that it can field in a particular phase. Empirically, this is of course another balancing act, the enduring norm here being parties' relative success in optimising *both* battle's soft factors (purpose, strength of character, preparedness, discipline) with its hard factors (hardware assets, force posture and availability of materiel) in order to win the fight.

Norms around defending and attacking are therefore unsurprisingly context dependent, but they are of course impacted by the set of circumstances immediately unfolding in front of troops. A new event or a new trend that threatens to upend current practice will straightaway

---

[15] Alex Horton and others, 'On the Battlefield, Ukraine Uses Soviet Era Weapons Against Russia', *Washington Post*, 20 March 2022, https://www.washingtonpost.com/world/2022/04/29/urkaine-russian-soviet-weapons/.

lead to provisional changes in that practice, the more so given shifts in general battlecraft, whereby concealment is ever more challenging, almost all movement now betrays positions and a force's activities now creates electronic signatures that can be readily identified and which give rise to immediate new angles of attack. All of this, moreover, can now take place at significantly greater ranges than in the past.

Nevertheless, a generalisation might also be that defenders now hold more cards than the forces attacking their positions. Defenders disproportionately benefit from, for instance, the saturation of battlespace by sensors. Surprising defenders and their positions is increasingly difficult just as it may be easier to identify adversarial activity in advance of an attack. Actually, the passing norm remains that both attacking *and* defending forces can be surveilled, detected and engaged by the opposing force almost regardless of that force's posture and it is therefore the *degree* of advantage and the duration of each new means (or, in the case of the defence, the success with which new means can be nullified) that may or may not convert any provisional adjustment in practices into a developing new norm.

In this vein, the Ukrainian conflict has highlighted the importance to unit survival of defeating an opposing forces' precision munitions (whether in offence or defence), a new priority therefore being the location of launch sites and platforms that are able to carry out such strikes. Precision artillery, moreover, is not only more effective than unguided munitions but also lessens force vulnerability through reducing that party's overall logistical footprint. As usual, versos to norm change exist, in this case that precision munitions will remain expensive, difficult to replace, more testing to use and susceptible to countermeasures from a prepared, well-resourced adversary. Nevertheless, parties' telegraphed rush to procure precision fires in light of Ukraine's experience evidences the importance now credited to the weapon type.

Tactics in Ukraine over the summer of 2023 remain characterised by an offence that relies upon occasional foraying but basically still framed by tactics of traditional attrition pitched against a defender that is now occupying prepared positions which are organised in depth. Indeed, intense combat has produced little change in the territories that respective sides control regardless of technology deployed. This has various explanations. Active fronts are long and deep and defenders have undertaken massive construction of defensive lines (the sowing of mines, the preparation of interlocking enfilade and defilade positions, the building of layered cover).

More importantly, battlespace is basically situational; today's defence can very quickly morph into tomorrow's offence in the same way that it is not particularly useful to label particular action types (be they manoeuvrist or positional, or the destruction wrought by attrition). Forces' new dispersion, moreover, make for a diffused set of targets that dilutes the effect and use of precision munitions. Here, the availability of precision still remains patchy. Smart munitions cannot be endlessly resupplied across a whole front, a small remaining comfort to the deployment of battlefield's traditional assets (tanks, personnel carriers, piloted aircraft) and their survivability.

Properly planned, resourced and led, the offence is of course not dead. Ukraine's September 2022 counter-offensive captured more than 6,000≈kilometres of previously Russian-held territory and did this in less than a fortnight with armour playing a considerable role. Indeed, notions of the breakthrough and the defeat of defending forces are still as valid as those underpinning Germany's conquest of France in 1940, Israel's progress in the Six Day War of 1967 or America's win in Operation Desert Storm in 1991. Despite poor execution, it must be remembered that Russia's initial invasion still gained more than 100,000 square kilometres of its neighbour's ground in less than a month, followed by Ukraine retaking more than half of that area in the subsequent two months.

Battlefield fortunes therefore remain susceptible to change and quick adjustment. The question is thus the *degree* of such tuning and the triggers that make provisional adjustment properly permanent given that this can still come about in all manner of ways, both military (a breakthrough or the sudden success of new means) and political (President Putin's decision in 2022 to enact a politically risky partial mobilisation, the actions of coalition partners securing change in behaviours). Indeed, norms should still reflect that to-and-fro in adversaries' fortunes is rarely tied to new means in warfighting. In this vein, stalemate in Ukraine in 2023 has occurred regardless of technology's new precision that promised to change norms so dramatically. Norms should still factor, for instance, that the armoured tank, when its deployment is contextually appropriate and is undertaken as part of a resourced, joined-up plan, will continue to play a material battlefield role. Indeed, the modern tank is vastly smarter than even its recent predecessors. It has better protection (reactive and often spaced armour), better control systems and better means of delivering lethality.

As noted by Biddle, the modern history of all land warfare is remarkably constant and, as such, should constitute an enduring norm: 'Since at least 1917 it has been very hard to breakthrough properly supplied defences that are disposed in depth, supported by operational reserves, and prepared with forward positions that are covered and concealed.'[16] It is little surprise that dug-in, prepared and well-led battlecraft less resembles blitzkrieg and looks more akin to grinding attrition. Indeed, Russia's experience of rapid early gains followed by its adversary's successful counterattack against overextended lines is far more similar than different to past practice. And while there is of course a range of new military equipment in Ukraine, *every* war brings new and potentially disruptive kit to the battlefield. It should not shock that its latest guise has yet to alter what is the unchanging nature of war. The central issue is unchanged: How can one side progress on a battlefield that has been prepared and exploited for every available advantage? The verso similarly remains that the properly resourced offensive persists as an enduring practice, the more so against shallow, ill-prepared defences supported by unmotivated troops and this characteristic is unlikely to be rendered obsolete by new technology.

Alliances, Coalitions and Norms

The greater available means for parties to wage war discussed throughout this primer highlights the importance of strategic *flexibility* and how norms must factor for this elasticity. The intricacy here arises from the awkward nexus in how norms reflect but also dictate passing practice. Indeed, planners and politicians alike must acknowledge that difficult edge where norms become an anchor to necessary change, whether from an alliance partner or domestic pressure group with opposing views and agenda. The role and purpose of norms should not be to encourage inactivity. This elasticity is therefore a product of both timeliness (particularly in decision-making) and imagination if a party is to compete successfully in this evolving meld of conventional but also hybrid means. While not at all a new norm, commanders can certainly not expect a neat battlefield and, in this vein, access to a theatre (here, a conflict zone) and assistance when in that theatre rely more than ever on a party's alliances and coalitions.

The convention remains that alliances secure participating parties both support and collaboration. Alliances bring teamwork. They backfill for weaknesses in one's own portfolio of means. They (as well perhaps as the

---

[16] Stephen Biddle, 'Ukraine and the Future of Offensive Maneuver', *War on the Rocks*, 22 November 2022, https://warontherocks.com/2022/11/ukraine-and-the-future-of-offensive-maneuver.

norms that they create) also provide an anchor that delivers stability against the volatility of those same domestic politics and, as such, a brake against reflexive change in participating parties' affairs and relationships. The enduring norm, however, is that alliances and coalitions all remain entirely political arrangements and ultimately subject to the febrility that political whim suggests. Indeed, given that it is the thousand small adaptions that can quickly trigger upheaval across practices, it is the *aggregation* of change agents identified in this book that shape norms for the timeline of this primer: societal transformations; the frequent precariousness of governments, their partners and institutions; new means of warfare but also degrees of overarching inertia that blunt those agents' ability to change and alliances that will require ever more attention if they are not to unravel or be subverted.

Alliances have other behavioural spinoffs that are relevant to passing norms. First (and concerning the West's matrix of arrangements), they focus participants upon the obligations of the West's role as a collective democracy. They reinforce the norm that Western operations must be morally acceptable, proportional, fully costed and funded. And demonstrably so. Indeed, this is simply a restatement on an enduring set of behaviours that has governed Western alliances since 1945, the norm that Western actions and decision-making must be accompanied by strong political oversight and a high degree of 'openness' that properly accounts to its constituencies. And, while this is not a new norm, the Ukraine experience has demonstrated that the ballast of alliances can help mitigate war's general frictions. While winning certainly matters, a developing norm is *how* victory is achieved and *how* then any subsequent victory is framed. An example here might be the West's own use of misinformation, disinformation, propaganda and deception, where and when such tactics can be appropriate and the degree to which they are likely to require more supervision than might be the case with illiberal adversaries.

This divergence is important precisely because it informs the degree of strategic flexibility that those in charge can deploy in times of fast-moving geo-politics once conflict is underway. Unlike in autocracies, an enduring norm is that Western leaders are largely regulated by very well-informed electorates and by a well-developed and sophisticated press. Nevertheless, while the autocrat's decision-making can be centrally directed by a leader who faces neither oversight nor accountability, cause and effect here are unclear, a consequence of Russia's action being, for instance, a palpable revival of NATO's purpose and the importance of collaboration and collective decision-making in how the West defends itself. Instead, Russia's norm-breaking actions have occasioned a *norm-affirming* response to those on the other side of the Ukraine conflict that was unforeseen by those ordering invasion. Finland and Sweden's intention to join NATO provides a case in point proving that momentous actions (which in the geo-political arena are almost by definition likely to be contentious) can still have unambiguous or foreseen consequences.

These may be reflected in norms but only after a period that considers the further compensating actions of others, the unforeseen responses here of Helsinki and Stockholm. And while not predicted, they will, over time, materially change norms as force equations around NATO's northeastern flank are transformed. Indeed, just as war sharpens the nature of alliances and coalitions.[17] This can be seen in Russia's courting of Iran, China and North Korea and the West's engagement with Ukraine's Central European neighbours. But so too does it lead to new refinement and tuning in the scope of these arrangements including the development of important non-war and often economic strategies. Evidence here includes a web of relationships that has developed between the Kremlin and a whole new supply chain to help in its military manufacturing efforts. It also includes Russia's relationship with the Wagner Group as its proxy for operations that may be unconscionable to Western democracies but appealing to near-peer autocracies.[18]

In judging norms, it is also useful to consider the *mix* of alliances and other partnership arrangements in the security space. Alliances are clearly a long-established norm with origins in the mists of time. They may be covert, but the trend today (and an established norm) is for them to be reasonably transparent in order to add weight to parties' deterrence. For this reason, the war in Ukraine has refocused members of the NATO alliance, its cohesion and the reasons for the alliance's existence, returning it to an 'alliance of necessity' rather than one of choice, the situation it found itself in after the end of the Cold War.[19] A developing norm might be that NATO's basis (a consensus organisation with many members and where decisions are taken unanimously) has provided new momentum to

---

[17] Rosella Cappella Zielinski and Ryan Grauer, 'Understanding Battlefield Coalitions', *Journal of Strategic Studies*, Vol. 45, no. 2, 177-185, 28 January 2022, https://www.tandfonline.com/doi/pdf/10.1080/01402390.2021.2011231.

[18] Ori Swed and Daniel Burland, 'The Global Expansion of PMSCs: Trends, Opportunities and Risks', OHCHR.org, undated, https://www.ohchr.org/sites/default/files/Documents/Issues/Mercenaries/WG/ImmigrationAndBorder/swed-burland-submission.pdf.

[19] Akshan de Alwis, 'A New Age of Multilateralism: Potential Solutions for the South China Sea Conundrum', *Diplomatic Courier*, 7 June 2016.

such structures with smaller groups of countries that share similar interests but now re-energised to enter into military partnerships. These represent foundational but difficult shifts, the delaying by Turkey and Hungary of Sweden and Finland's accession into NATO evidencing the difficult of aligning interests across competing parties, across large geographical distances and accounting for cultural anomalies.

The same holds true in the case of the United Nations. The UN's founding coincided with the creation of similar large multilateral organisations such as the General Agreement on Tariffs and Trade in 1947. Today, an effect for norms might be that the 'ideal vision of global cooperation now stands compromised'.[20] Frictions and tensions abound, both on and away from the battlefield. Cross-border institutions may have failed to maintain cohesion in the face of new global hotspots and conflict in a manner similar to multilateralism being challenged by a pandemic despite the World Health Organization's best efforts. The norm is that countries choose their preferred partners and strategies with an eye on what everyone else is doing. In the place, then, of multilateral organisations, 'minilaterals' have sprung up, alliances based upon targeted initiatives by a handful of states to address a specific issue. The developing norm might therefore be that 'formal organizations persist, but governments increasingly participate in a bewildering array of flexible networks whose membership varies based on situational interests, shared values, or relevant capabilities'.[21] An example is provided by the Quadrilateral Security Dialogue composed of Australia, India, Japan and the United States. Whilst not a formal security alliance, it is focused on countering Chinese influence. Its small size allows speed, flexibility and agility in its responses.

The war in Ukraine might have tightened NATO's coherence but, with 30 countries in its cohort, it still only represents 15 per cent of the world's nations. The norm remains that several African and Latin American countries reject Western sensibilities about the Ukraine conflict just as other major states have remain non-aligned. The UN General Assembly's vote in April 2022 around Russian violations of human rights, highlights the empirics of this division; while the resolution achieved 93 votes, 24 were cast against its adoption with 58 parties abstaining. Abstentions involved Brazil, Belize and El Salvador in Latin America; Angola, Nigeria and Niger in Africa;

---

[20] Aarshi Tirkey, 'Minilateralism: Weighing the Prospects for Cooperation and Governance', Observer Research Foundation, 1 September 2021.

[21] Stewart Patrick, 'The New "New Multilateralism": Minilateral Cooperation, But at What Cost?', 18 December 2015, *Global Summitry*, Vol. 1, No. 2, pp. 115-134.

and Bahrain, Saudi Arabia and Qatar in the Middle East. Breakdowns in cohesion are not new and often baffle Western commentators.[22] However, they increasingly posit a change in norms. A case in point may be provided by the formal Non-Aligned Movement (NAM) as the largest grouping of states aside from the UN.

Smaller groupings also represent a more practical approach to alliance building. They maintain a useful distance between national needs and those of broader, more wide-reaching organisations. Regional security groupings have also had success such as the UK-led Joint Expeditionary Force (JEF) composed of just eight NATO members as well as Finland and Sweden. Founded in 2015, it follows NATO's Framework Nations Concept where a central power helps coordinate a more regionally focused group of nations. The JEF is focused on the defence and security of Northern Europe.[23] Importantly, the JEF is not a consensus organisation, and instead operates under a 'UK plus one' structure whereby the UK can conduct JEF activity alongside one or more JEF participants. There is no obligation to contribute forces by any one or more JEF participants.[24] The JEF complements NATO structures. Growing numbers of groupings can be difficult to understand and likely cause some confusion. Military cooperation is not simple, and even integration within nations can be difficult. Appropriate levels of interoperability require forces to train together extensively to understand each other's way of operating and nuances of deployment. Language and process differences must be worked through.

The defence industrial capacity has similarly had a long history of small group cooperation. The Eurofighter Typhoon fighter jet, for instance, was created by a consortium consisting of Germany, Italy, Spain and the UK. Now, the UK, Italy and Japan are cooperating over the development of a next-generation fighter under the Global Combat Air Programme. Indeed, industrial cooperation has unsurprisingly returned to the fore during the conflict in Ukraine, with states realising that their ability to produce munitions at scale has drastically weakened.[25] The entirety of Europe

---

[22] Elizabeth Sidiropoulos, 'How do Global South politics of non-alignment and solidarity explain South Africa's position on Ukraine?', Brookings, 2 August 2022.
[23] Sean Monaghan and Ed Arnold, 'Indispensable: NATO's Framework Nations Concept beyond Madrid', Center for Strategic and International Studies, 27 June 2022.
[24] UK Ministry of Defence, 'Joint Expeditionary Force (JEF) – Policy Direction', Policy Paper, 12 July 2021, https://www.gov.uk/government/publications/joint-expeditionary-force-policy-direction-july-2021/joint-expeditionary-force-jef-policy-direction.
[25] Sam Cranny-Evans, 'Ramping Up: What Will It Take to Boost the UK's Magazine Depth?', Royal United Services Institute, 6 December 2022.

now has just two factories with the appropriate machine tooling to make tank barrels. Rebuilding an industrial base with the suitable equipment and expertise is a very long-dated exercise, but also one where industrial partnerships may be as important as military ones. The broad norm remains that partnerships and alliances must be pragmatic. Parties will mix and match the type of partnership depending on need, even sharing bilateral and multilateral relationships with the same states.

Resistance and Norms

Discussion of means must also consider the role of resistance in warfighting. Partisan and insurgent warfare has long imposed costs on occupiers and will continue to prevent invaders from consolidating gains. Indeed, the developing norm here is that the same expanded toolkit of means is now available to aggressors and resisters alike. Here, the Ukraine conflict demonstrates that resistance covers a widening portfolio of means, many of which are accelerated by that same set of drivers around context, narrative, protagonists' understanding of their own history and, of course, by political will and its leadership. More granularly, norms here are shaped by local matters, by provincial constraints and special interests, by alliance structures as well as the social dynamics of each invaded party.

Recent advances in communications and weaponry would appear to facilitate the role of the resister, creating more and lower cost means to exact cost from an invader in an effort to change the aggressor's military decision calculus.[26] The norm remains that such campaigns must involve violent resistance and civil confrontation and to do this in tandem, leveraging those same information tools and technologies that interleave this primer's analysis. Just as a well-coordinated sabotage effort can wreak havoc on a larger superior invading force, technology can also enable the hidden, highly skilled few whose effect here can be considerable.[27] The verso remains that there are few empirical cases of resistances, by themselves, defeating the more powerful occupier.

---

[26] Economist editorial, 'Ukraine's Partisans Are Hitting Russian Soldiers Behind Their Own Lines', *Economist*, 5 June 2022, https://www.economist.com/europe/2022/06/05/ukraines-partisans-are-hitting-russian-soldiers-behind-their-own-lines.

[27] Peter Beaumont and Isobel Koshiw, '"The occupier should never feel safe"; Rise in partisan attacks in Ukraine', *Guardian*, 6 June 2022, https://www.theguardian.com/world/2022/jun/06/ukrainian-partisan-attacks-surge-russia.

Setting Priorities and Their Effect on Norms

Ukraine's effect on *others'* military priority-setting is not yet clear. It is, after all, complicated both at the strategic (the resurgence, for instance, of NATO and shifting burdens of responsibility created by changes in alliances[28]) and tactical levels (weapon availability and allocation, issues of resupply, stocking levels and logistics). Here, the evolving norm is shaped by a new urgency to secure supply chains, to ensure better and broader states of readiness, superior resilience and a binding together of parties' constituencies. It must also factor for increased stockpiles and, more generally, an increased buffer to operations. Priority-setting, moreover, is no less complicated in an age of sophisticated planning tools and machine-assisted system management. Logistics may no longer involve a quill pen but planners instead have to contend with more bits from more suppliers in more geographies, all working to different durations.

Other factors complicate the life of planner and commander, not least the persistent disarray that is intentionally sown by the liminal, sub-threshold activities of adversaries. Ambiguity is an enduring norm and the convention remains that war planners must continue to factor for the *cumulative* effect of their schemes rather than outcomes achieved from individual actions. Threats to planning, moreover, remain the strategic misstep and the erroneous crossing of boundaries between war and non-war. In this vein, a further norm dates from the early 2000s concerning the setting up and use of parties' de-escalation channels.[29] With fewer structures now available for parties to mediate or intervene, an emerging norm might be for ratchet-like increases in geo-political tension, the more so given parties' competing informational narratives that reduce clarity around actors' involvement, objectives and operations.

This therefore points to *additional* instability in behaviours over the period under consideration by this primer. Indeed, Ukraine has already highlighted the risks that are inherent in lowering the barrier to conventional conflict between major powers given that certain participating parties have nuclear weapons. How indeed can a nuclear power lose a conventional war? Here, Russia's sabre-rattling around its nuclear strategy has seeded

---

[28] Adam Tooze, 'The Second Coming of NATO', *New Statesman*, 18 May 2022, https://www.newstatesman.com/international-politics/geopolitics/2022/05/the-second-coming-of-nato.
[29] Forrest Morgan and others, *Dangerous Thresholds: Managing Escalation in the Twenty First Century*, RAND, 2008, https://www.rand.org/content/dam/rand/pubs/monographs/2008/RAND_MG614.pdf.

new doubts in how *conventional* deterrence can ever again be achieved. This will certainly affect norms but it also raises questions about how strategic priorities can be reset post-Ukraine without nuclear weapons being part of that new calculus.[30] The effect, of course, of interfering with such previous ambiguity around nuclear capabilities and intentions can only destabilise current structure and, by extension, passing norms.

A further long-dated factor influences priority-setting, this time around parties' perennially scarce resources and how to allocate shrinking budgets, whether in strategic and absolute terms (adjudicating funding priorities, funding long-dated commitments of a new weapons programme) or, more tactically, the assignment of spending between, say, conventional and then grey zone activities. Two observations arise. Ukraine has demonstrated that overt provision by its allies of military materiel risks both escalation and the number of attack surfaces for unforeseen confrontation.[31] Second, history reminds planners (especially in dynamic and fast-moving environments) that ill-judged investment in particular verticals or inappropriate projects risks creating obsolete and orphan assets that are unlikely to generate expected impact.

A challenging realisation for middle-order parties might therefore be that the relevance and fit of trophy systems is collapsing. It seems inconceivable that the assumptions under which the UK has recently deployed its new carrier fleet will be relevant in 2030 let alone by 2060, the ships' planned retirement. This is not to suggest that planning will become an easier exercise. While hybrid means may appear to save money on the portfolio of expensive assets required to wage conventional war, the prevailing norm suggests otherwise: Hybrid effects are difficult to forecast and it remains that Russia's invasion of Ukraine highlights again the need to prepare for high intensity operations rather than the skirmish activities associated with asymmetric enemies.

---

[30] Rebecca Johnson, 'Ukraine war shows nuclear deterrence doesn't work. We need disarmament', Open Democracy, 24 March 2022, https://www.opendemocracy.net/en/odr/ukraine-russia-war-putin-nuclear-weapons-disarmament-deterrence/.

[31] Guy Faulconbridge, 'Russia warns US against sending more arms to Ukraine', *Reuters*, 25 April 2022, https://www.reuters.com/world/europe/russia-warned-united-states-against-sending-more-arms-ukraine-2022-04-25/.

5
Acquisition and Integration of Novel Systems into Legacy Force Design

Novel technologies present their own set of challenges in considering norms. Claims and deployment assumptions made for these systems' procurement have frequently proved plain wrong, and usually disappointing relative to projects' initial cost and performance forecasts. The capabilities of novel weaponry have routinely been overstated since time immemorial and then extrapolated by politicians and military staff alike to create transformative narratives without proper regard for the challenges posed by those systems' delivery, their deployment and integration. Indeed, the convolution involved in introducing new weapons to the battlefield hides several (and usually underappreciated) points of friction.[1] Delivery is often late, prone to political interference, complicated by the multiplicity of involved parties and, over the long procurement timeline, a general fraying in accountability. Systems appear unvaryingly overbudget and are then upended by technology, either by subsequent technical developments or by unsolvable technical difficulties that appear at several points in time ahead of delivery. Moreover, once signed off by decision-makers, these same technologies often gain inappropriate and gravitational power that

---

[1] Lucia Retter and others, *Persistent Challenges in UK Defence Equipment Acquisition*, RAND, 2021, https://www.rand.org/pubs/research_reports/RRA1174-1.html.

then stifles meaningful subsequent debate about the how, the when and the where new systems should be deployed.

Procurement and Norms

At one end of the argument, the scale to which transformative technologies are expected to be disruptive rarely accords with empirics. Assumptions seldom account sufficiently for the twin challenges of cost and complexity. At the other, intricate systems then fall short of being the intended comprehensive replacement for a retiring platform. They are often incremental, their eventual deployment elegantly masking the underlying problems that had already bedevilled existing systems. In this vein, the norm remains that military procurement rarely offers a silver bullet for a party's embedded shortcomings, be they technical, doctrinal or simply arising from a party's dwindling lack of mass. There is also little empirical correlation between a state's delivery of hardware and that state's overall security needs which, after all, depend upon a complex portfolio of factors that include the number, delivery, effect and readiness of these weapons as well as their ease of integration with legacy system. Does a party's portfolio perform better once that novel platform is deployed? Has deterrence been improved, the notion of winning the war before it starts or, at the very least, preventing an adversary's deployment?

Planners' investment in novel systems is driven by a complicated equation of factors. It is shaped by debate around *new* domains, for instance around cyber and space as the new frontiers of warfare, and the degree to which existing land, air and sea platforms remain relevant and fit for purpose given changes in adversaries' force posture and capabilities.[2] A second norm arises. Procurement is not generally a question of substitution but one of amelioration, of upgrading, amalgamation and integration. Indeed, acquisition and integration of novel systems are complicated precisely because of the volatile nature of relationship between new technology and the maintenance of current operations, between available assets and passing doctrine, between available budgets and the cost of new programmes and, latterly, a new requirement for combat mass in future operations.

---

[2] UK Thoughts on Defence, 'Retiring Sunset Capabilities in the Integrated Review', onukdefence.co.uk,   12   March   2021,   https://onukdefence.co.uk/military-capability/retiring-sunset-capabilities-in-the-integrated-review-you-have-to-trust-someone.

Second, it is governed by connections that must exist between incoming hardware, current practices and the available cohort of trained (and training) operators available to settle new kit into existing arrangements. All of this requires intricate coordination in order, notwithstanding such multiple (and often subjective) variables, that programmes be tested, validated and then seamlessly integrated across the whole heterogenous force. The verso here is that procurement takes an age, is very process-heavy and involves considerable 'technical debt' (the consequences of poor design, changing architectures, commercial pressures, the difficulty of testing combat assets in peacetime environment, the eventual merging of procurement pieces into a deliverable product and, lastly, the trials of configuring new assets to account for the often-disparate service priorities of receiving parties). Moreover, acquisition practices must factor that developments in one domain require lock-step advances in others if they are to translate into proper effect.

None of this is new and none of this yet adds up to norm change. In particular, it is data's presumed underpinning of battlespace that has ramifications on norms. Notwithstanding novel systems' broad reliance upon machine workings, the challenges of a contested electro-magnetic spectrum are recognised but remain poorly prioritised. Data is pivotal and a likely pinch point in novel system's ability to deliver expected performance. The effects, after all, on operations of partial, duplicatory, obsolete and contradictory data must properly be factored with feint, spoofing, disinformation and other adversarial measures in the deployment of new means. Procurement of smart kit that can readily be rendered dumb means that soldiers must be prepared still to fight with their knuckles in order to win the battle.

The practice therefore remains that assets in space, in the electronic ether or upon remote platforms may represent the new forms of warfare but cannot yet be framed by a reliable set of norms of their own. Instead, their deployment must be undertaken on the conservative basis whereby users should at least expect to fight blind, without the data and connectivity that underpin their operation. Weapon advantage, moreover, depends upon its own portfolio of often intangible factors, upon reliability, on easy employment and buy-in from its users, on maintenance that is both minimal and practical, on upgrades and seamless integration with colleague assets, on flexible configuration and modularity, on resistance to adversarial meddling and, again, on the appropriateness for the task in hand. It also depends on an understanding of how adversaries are operating, projected frailties of their systems and, as above, an acceptance that operators must train for their own systems' degradation in hot use. These may appear long-dated factors but they also appear far removed from the agile, responsive procurement of kit that sets current headlines.

In considering how novel systems will affect norms, it is useful to construct scenarios, to role play and then to think about the issue through this lens. Swarming and loitering munitions are a case in point and one that promises an interesting discontinuity in reach and application to battlefield commanders. Understanding likely pitfalls in swarm deployment might therefore provide transferable context to the degree of norm change possible from adopting adjacent new technologies over the two decades considered by this primer. Here, uncrewed systems require seamless data links. In a communications-denied environment, these munitions will certainly need to backfill for that same partial, incomplete, or hyper-sensitised data set out above. They will in time need to do this without a human in the loop. They will require resilience and on-platform routines to automate, manage and optimise performance. They will need expert configuration (boundaries, permissions, triggers, fail protocols and elements of an overarching value system) and all of this notwithstanding that their tasking will be problematic given the dynamic nature of both battlespace and colleague assets (both supervised and unsupervised).

While sensors may be able to provide them a degree of situational awareness, swarm technology must include priority setting, feedback loops to monitor and communicate performance as well as processes to set goals (and to do this in line with their commander's values and in accordance with the rule of law and local rules of engagement). Swarm systems should include procedures to undertake attribution and forward planning. In the particular case of their targeting capabilities, swarm deployment will still be complicated by terrain considerations, the moving parts of a battlefield, by camouflage and decoys, and by the need to reorganise after contact with the enemy's point defence systems. Unsupervised weaponry also requires that targets be labelled, classified and allocated in real time if these systems are to be empirically dependable and accretive to their users. Remote platforms, moreover, must still be subject to the same restrictions on size, performance and range, the same weight, cost and stealth constraints as their crewed alternatives. They must communicate changes in state, 'understand' and own that state in relation to other (and often legacy) assets on the battlefield as well as the effects of their interventions. These, then, are the component parts that will comprise successful deployment of such systems.

The point here is that integrating novel systems into legacy force design requires complicated trade-offs. The factors are nuanced and concern initial configuration, their calibration and maintenance, updating and storage; their defence capabilities; and their recovery, power and survivability as well as integration of their tasking into wider war aims. They must work first time and every time and all of these issues require transparent resolution in order for the platform to be effective. These machines need also to be cost efficient, replaceable and usable by operators with transferable skills and, in instances of malfunction, the systems should be able to rely upon embedded previous 'experience' in order to complete given tasks. The pathway between deployment, integration and normal change is therefore unlikely to be straightforward. It is these frictions that also inform the degree and type of norm change in this matter.

Procurement Frictions

The passage of the CHIPS and Science Act in the US in 2022 evidences the importance of norms in shaping procurement and the techno-security framework that governs acquisition of military assets.[3] Geo-politics, domestic politics and political positioning all influence procurement notwithstanding the ground rule that the acquisition of new forms of warfare should generally be determined by the activities of adversaries and by the prerequisite for governments that the country is kept safe and able to meet its security obligations. Complexity arises from the mismatch between politicians' reflexive response to an adversary (China, for instance, is notably upgrading available means to achieve better than parity with likely challengers) and the passage of time that exists between concept and delivery of weapon systems.

Nor can procurement be undertaken in isolation. In the case of China, its aspiration requires a patient programme which includes the coordination of ambitious and long-term science, technology and innovation plans.[4] Indeed, the enduring norm is that procurement is generally a *national* endeavour that comes to define national priorities and the allocation of

---

[3] See Justin Badlam et al., 'The CHIPS and Science Act: Here's what's in it', McKinsey & Partners, 4 October 2022, https://www.mckinsey.com/industries/public-sector/our-insights/the-chips-and-science-act-heres-whats-in-it.

[4] Tai Ming Cheung and Thomas Mahnken, 'The Grand Race for Techno-Security Leadership', *War on the Rocks*, 31 August 2022, https://warontherocks.com/2022/08/the-grand-race-for-techno-security-leadership/.

resources. Its prominence may appear to ebb and flow but this is largely irrelevant given procurement's very long timelines. Its drivers may also appear to vacillate (Cheung and Mahnken note, for instance, that the 'scale, pace and cost [of] this ratcheting up of efforts by Washington and Beijing looks set to far [outpace] what took place between the United States and Soviet Union') but these are often set by governments having to match procurement policy with domestic audiences as much as aligning to strategic considerations. Procurement, after all, is a component of strategic competition and, while the norm may look unchanged from the Cold War era, the mechanics under which this rivalry currently plays out are dynamic and shifting. Its systemic importance is even evidenced by the manner in which countries are organised from China's state-led, top-down approach and the market-driven, bottom-up US system.

Procurement and acquisition programmes are therefore driven by a set of overarching foundational norms (state security, permanent competition, fulfilment of useful alliance commitments, maintenance of advantages, leveraging others' weaknesses, demands arising from political and force posture, projection of power) from which behaviours are subsequently driven. Three primary observations arise from this framework (East versus West, China versus America, democratic versus illiberal parties). First, it moves slowly, even at a generational pace. It is quite well telegraphed. It is also a mechanism that seeds a host of tactical and incidental behaviours along the way, each of which may have deep consequence, have shorter duration, and may or may not then make the transition from evolving to new norm. That China, for instance, has been acquiring new weapons 'five times faster' than its Western adversaries must inform norms, not least because earlier assumptions of China being a struggling technological laggard are clearly outdated. Similarly, any degree of technical complacency within Western procurement bodies has been replaced by an admission that their own capabilities require overhaul if they are to match the Chinese resurgence. Third, norm development arising from this procurement framework spreads *across* borders and is picked up and copied by competing states including, in the case of China, Beijing's new arrangements to spur innovation through military-civil cooperation and self-reliance, a significant amendment to a traditional dependence upon central planning systems and state-determined priorities and allocation.

Developments in procurement also drive *behavioural* change and this may not always be constructive. In the US, for instance, public shortcomings and examples of egregious conduct mean that the defence industry is unhelpfully lumped together as a single suspicious entity and held at arm's length. Contrary to the Cold War period, the current norm around America's public-private procurement relationship is one of dubiety and distrust. This has important normative consequences; in the case of America, the perception that its acquisition processes appear more and not less rigid, more and not less risk-averse has led to the accusation that the vertical is insufficiently innovative.

'Technical Debt' in Weapon Procurement

The perception persists, moreover, that defence acquisition is inefficient.[5] But complex systems require complex manufacturing processes and, in the case of military acquisition, procurement is subject to additional frictions from the configuration and subsequent integration of equipment (including the limited resources and conflicting priorities of those receiving parties), the mismatch between what is produced by a country's procurement executive and what is expected by the teeth arm that will use the kit, and, more importantly, complexities arising from the integration of this kit into existing legacy systems. These are behavioural, subjective challenges that require more than new software tools in order to manage the processes needed to integrate such programmes into existing structures and kit lists.[6]

An emerging norm then relates to the asymmetric costs of certain of these programmes (in the case of the UK, the budget required to fund the acquisition, integration and long-tail costs of its submarine, aircraft carrier and aircraft programmes) that have each become an increasingly disproportionate share of the country's overall defence budget.[7] These represent not only historically outsize commitments but also *longer* term commitments than have previously been undertaken, and all shouldered in the face of rapid development of possibly disruptive capabilities (artificial intelligence, predictive machine learning), all with the potential to upend current assumptions and planning cycles. Shorter production runs of

---

[5] Lucia Retter and others, *Persistent Challenges in UK Defence Equipment Acquisition*, RAND, 2021, https://www.rand.org/pubs/research_reports/RRA1174-1.html.
[6] Jhon Spellar, 'Smart procurement: an objective of the Strategic Defence Review', *RUSI Journal*, 1998, 143.
[7] UK Government, 'Defence Equipment Plan, 2022 to 2032', https://assets.publishing. service.gov.uk/government/uploads/system/uploads/attachment_data/file/1120332/ The_defence_equipment_plan_2022_to_2032.pdf.

'exquisite' novel systems also mean that production efficiencies will be harder to achieve, with norm change that would otherwise arise from experimentation and adaption being correspondingly muted.[8]

Integrating novel technology into organisations that have been traditionally suspicious of change makes that task harder. Such hesitancies, after all, have deep historic roots and can be justified on several levels. Forces' personnel have a job to do regardless of the kit with which they are provided. Over the span of a career, perhaps 25 years for those responsible on the ground for integrating new means, technology comes and goes. Behaviourally, those practitioners are much more concerned about the exertion involved working through issues and the technical debt that often accompany these projects (the costs arising from inappropriate design, workarounds, training colleagues, creating protocols in order to embed these new practices, their testing and validation and, in time, coordinating the withdrawal of superseded equipment as well as then 'training out' their use).

Unsurprisingly, therefore, acquisition programmes have long presented a demanding series of cliff faces and, given that these schedules have traditionally been measured over decades, the norm is unlikely to change over the period of this primer. There are many degrees of separation between personnel specifying, coordinating and delivering new systems and then those lucky individuals tasked with accepting, bedding in and then using these assets. Biases around 'not-invented-here' are difficult to remedy. Three issues reinforce the point. First, novel platforms often risk obsolescence at this very point of delivery given the long timelines required to create these systems, to iron out their configuration issues, to overcome integration inertia and to do all of this while factoring for the fast-changing character of war. Second, the variability of that change across sectors, geographies and categories complicates integration procedures.[9] It is rare that programmes respond to a one-size-fits-all approach with each new project requiring re-learning and different proficiencies. A third matter is again behavioural and is rooted in how forces have traditionally procured

---

[8] Eric Tegler, 'Russia may be showing it is running low on precision guided munitions', *Forbes*, 24 March 2022, https://www.forbes.com/sites/erictegler/2022/03/24/from-debuting-hypersonic-missiles-in-ukraine-to-hinting-at-chemical-weapons-russia-may-be-signaling-its-short-of-munitions/?sh=21ba7480632a.

[9] Olivia White and others, 'War in Ukraine: 12 disruptions changing the world', McKinsey and Partners, 9 May 2022, https://www.mckinsey.com/business-functions/strategy-and-corporate-finance/our-insights/war-in-ukraine-twelve-disruptions-changing-the-world.

their capabilities. Inter-service rivalry is still a factor in procurement. So is traditionally siloed thinking that still persists between Navy, Army and Air Force, the more so given that tactics and doctrine, themselves additional sources of friction, are key enablers in these systems' successful deployment across arms and services.

Contradictions also arise from the legacy equipment and practices that remain in states' arsenals and the notion that new replacement equipment has been procured to address requirements but also *shortfalls* in that party's capabilities. These, after all, may be deep-seated and date back to earlier generations. This is not helped by the febrile set of drivers that often characterises procurement (government interference and point-scoring, changing personnel, muddled responsibilities, opaque commercial priorities). The norm remains that new platforms are never specified, procured and deployed unencumbered. Furthermore, the erratic pace of innovation and the random traction that involves military equipment (today's purported 'revolution in military affairs' is often tomorrow's old news) tends, as a generalisation, to frustrate development in 'use' norms. Finally to this point, better transparency in procurement may heighten parties' fear of missing out on that new disruptive technology.

There is always, therefore, a danger that norms around new capabilities are highjacked by exogenous forces, be they by politicians, special interest groups within the military family or, indeed, by the media. The press regularly reports, for instance, that the UK's command and control capabilities, fundamental to Western battlecraft, are moments away from being degraded by adversarial deployment of new means. This elicits kneejerk and often ill-informed efforts at remediation which at the very least influence short-term allocations (if not actual purchasing plans). The phenomenon is especially relevant in the newer domains of warfare where investment in space and cyber assets promises outsize disruption relative to traditional battlefield assets, the more so given the lead by commercial parties coming up with dual-use technologies and doing this at unprecedented speed.

Contradictions therefore abound, not least as Western militaries' pivot from counter-insurgency to conventional operations and the degree of obsolescence that this has already created in parties' equipment and practices. While Ukraine has also demonstrated that platform design and doctrine-for-use require similar change in order to be fit for use, this also does not equate to a new norm. Two observations arise. The empiric here for the UK has been that 'change is the only constant' in this field and, second, that political expediency and inadequate investment has long led to a reduction in the warfighting capabilities of the British Army. In this vein, the requirement to 'adapt and leverage remaining advantages' which underpins the 2021 Integrated Review is not in itself a fresh clarion call but instead the repeat of an often-stated refrain.

A further facet in this notion of 'technical debt' concerns the utility of old equipment in warfare. Ukraine has demonstrated that older weapons remain exceedingly lethal given appropriate deployment and support; a chunk of well-placed metal does not need to be a next-generation super weapon. The enduring norm is that considerable damage can be wrought by adhering to old practices. Russia evidences this in its resort to older generation missile technology to attrite Ukraine's energy substations despite its portfolio of offensive cyber capabilities and other novel equipment. The phenomenon even questions parties' logic in pursuing ever more technical and exquisite capabilities, the point at which development's costs and complexities outweigh promised advantages.[10]

Newer, fewer and more complicated platforms seem to have dominated Western procurement projects for decades, all with their own excruciating journey before they finally make it into service. Ukraine's priority to deploy systems that are 'good-enough', the primacy of mass and immediately deliverable means to fight the battle may refocus planners to 'break the connection between cost and improving... force effectiveness'.[11] UK examples of complex procurement projects currently undergoing integration into legacy practices include delivery of the F-35 Lightning fighter aircraft, its Ajax and Boxer armoured vehicles, its two aircraft carriers, the delivery of P-8 Orion maritime patrol aircraft and the Protector remotely piloted air system. Only small numbers of these expensive platforms can be procured. Highly integrated, these systems are also difficult to modify and iterate as time goes on. Indeed, during the period in which the US developed its F-22 Raptor fighter jet, the USSR and then Russia fielded six separate generations of air defence systems.

Technical debt is also created by high levels of attrition. It is often argued that the 'containable' losses of materiel experienced during the wars in Iraq

---

[10] Mykhaylo Zabrodskyi and others, 'Preliminary Lessons in Conventional Warfighting from Russia's Invasion of Ukraine: February-July 2022', Royal United Services Institute, December 2023, 56.

[11] Bryan Clark, Dan Patt and Harrison Schramm, 'Mosaic Warfare: Exploiting Artificial Intelligence and Autonomous Systems to Implement Decision-Centric Operations', Center for Strategic and Budgetary Assessments, 2020, 12.

and Afghanistan have mistakenly shaped current planning assumptions as well as their underlying heuristics. After all, the very different procurement environment of counter-insurgency operations is all that the majority of today's planning cohort has really known. Apocryphal evidence, however, suggests that the Ukrainian battlefield is already prompting planners to question the wisdom of buying a few, very capable weapon systems. Once they are gone, they are gone. While military uncrewed aerial systems (UAS) may be designed to fly for thousands of hours above the combat zone, experience from the front line has undone these assumptions. In the early stages of the war, the average life expectancy for a quadcopter was three flights, and for a fixed wing UAS it was around six flights.[12] To that end, there is no point in designing in sophisticated resilience and redundancy when the unit is only expected to fly a matter of hours and this should gradually change the norms under which planners make their decisions. A distinction is required between what is likely to survive and what will not. Single exquisite systems, moreover, should not be considered the answer for all eventualities, undermining the notion of the multi-mission platform that is capable (at some cost) of doing everything. Complex systems are harder to maintain, repair and upgrade. Deploying and, as appropriate, combining systems with more limited functionality might instead create sufficient effect on the battlefield. Mass, moreover, has its own (and often overlooked) utility. If system simplicity is to return as a procurement driver, this may allow defence processes to be similarly simplified and their timeframes shortened.

Norms and Uncrewed Combat Assets

The advent of small uncrewed aerial vehicles (UAVs) and their part in future battlecraft merit review as their role in Ukraine has demonstrated the significance of this new weaponry in shaping practice. Two characteristics are significant from the first year of this conflict as they suggest evolution in current norms. First, in its surveillance role, the observed are now less likely to know that they are being watched. Second, the array of capabilities facilitated by UAVs is surprisingly broad and still in its infancy. Uses for the platforms range from a means to gather intelligence to being a targeting

---

[12] Mykhaylo Zabrodskyi and others, 'Preliminary Lessons in Conventional Warfighting from Russia's Invasion of Ukraine: February-July 2022', Royal United Services Institute, December 2023, 37.

platform, a munition and, in time, a remote and possibly unsupervised colleague asset in its own right. The general targeting cycle, already less than a few minutes (and leading itself to a change of norm), is reducing quickly because of their deployment.[13] Nor is Ukraine the first conflict to showcase their capabilities, which were earlier demonstrated in 2020 in Nagorno-Karabakh, in 2018 in Syria, and through extensive deployment in Libya, Saudi Arabia and Iraq. Notwithstanding frailties (susceptibility to electronic and other countermeasures, requirement for detailed operator training, the empirics of very high attrition rates) the platform is already proven.

Recent conflict zones have therefore provided key testbeds to develop UAVs' role and subsequent procurement parameters including, for example, their deployment as a loitering munition.[14] The verso, however, is that these systems, more similar to a missile than an aircraft, remain just another battlefield consumable. Challenges still abound with the weapon type. Range is limited, reducing their use to the tactical scale. A UAV is susceptible to electronic countermeasures, requires skilled operatives and must be integrated into wider operations if their use is to be impactful. Usually a one-time-use weapon which, having located a target, mostly crashes into it to deliver lethality, its capabilities provide a developing new form of warfare and one where catch-up is already required in doctrine and training to achieve the means' proper integration into cross-domain practices. UAV operators, moreover, are no longer reliant on airfields or large open areas to deploy their means, an emerging norm being their broad deployment across the battlefield. While its effects differ based on model designation, a new convention here is that all adversaries must at least assume they are persistently under threat from the more capable versions available and must now plan their operations accordingly. No part of the battlefield is a safe area.

The advent of UAV loitering munitions also enables small, dismounted teams to replicate (at lower cost, with quicker engagement cycles and remote operation) the traditional role of artillery. Their deployment extends anti-armour operations well beyond line-of-sight distance (other artillery assets already fill this role) and threaten armour in pre-prepared defensive positions as well those exposed during manoeuvre. Notwithstanding

---

[13] Sam Cranny-Evans, 'As small drones shape how we fight, is the British army ready to face them?' Royal United Services Institute, 21 July 2022.

[14] Brennan Deveraux, 'Loitering Munitions Is Ukraine and Beyond', *War on the Rocks*, 22 April 2022, https://warontherocks.com/2022/04/loitering-munitions-in-ukraine-and-beyond.

supply constraints, UAVs' cost-precision-remote equation appears to herald a discontinuity, the more so given it remains early in its development cycle. Drones thus represent a new form in warfare that has already caused current norms to flex as operations morph to account for these assets, both in their deployment but also in the actions required to counter adversarial use. Assuming penetration in time of a party's advanced air-defence networks, UAVs will then be able to challenge the notion of the protective anti-access bubble (a key defence concept to the likes of China and Russia).

A further norm change from UAV deployment is that physical bases have become particularly vulnerable to drone attack, not least because there is poor delineation of roles and responsibilities in defending against such assault.[15] UAVs also work in concert with forces' current legacy practices and do not necessarily require wholesale changes in doctrine. They operate as an adjunct service which, given the fast-paced evolution of drone warfare, mitigates policymakers' difficulties over ensuring common strategic vision and practices for their deployment. Rather, several legacy systems will increasingly be neutered by UAV use. It is difficult to establish how successful either side in the Ukraine war has been at defeating attacks by uncrewed assets. A feature of massed drone attack is that just one or two weapons need to penetrate defences in order to create effect and seize headlines. In the Middle East, after all, Houthi attacks in the Gulf States have shown that legacy air defence systems are less and less effective in combating UAV threat, particularly against drone swarms with decentralised flight patterns.

A massed drone attack controlled by decentralised operators makes it challenging for defenders to neutralise each and every angle of attack. UAVs' size, slow speed and often plastic construction helps them avoid detection by current anti-aircraft sensors and, relatively inexpensive, the evolving norm here is that they should in time be readily procurable and deployed in large numbers, the notion of force multiplication. During the first half of 2023, more than 80 per cent of all Ukrainian targets were derived from drones.[16] With artificial intelligence facilitating better integration, an evolving norm must be that swarms are likely to become ever more lethal. Here, the Ukrainian conflict demonstrates the overall nature of challenge

---

[15] Nicholas Paul Pacheco, 'How Doctrine and Delineation Can Help Defeat Drones', *War on the Rocks*, 13 December 2022, https://warontherocks.com/2022/12/how-doctrine-and-delineation-can-help-defeat-drones.

[16] Economist Editorial, 'Ypres with AI', *Economist Special Report on Warfare After Ukraine*, 8 July 2023.

posed by drone assets with Iranian Shahed-136 kamikaze assets replacing long-range precision fires as Russia's preferred method of aerial assault.

The US Valkyrie programme similarly points to the shape of next generation drones and to how changes to these platforms might reflect themselves in norms over this primer's timeline.[17] Developments in the weapon class also reinforce the possibility of building fleets of smart but relatively inexpensive weapons that are deployable in large numbers and so upending current practices. The Valkyrie prototypes can already fly distances equal to the width of China, supplement the role and reach of existing fighter jets and provide human pilots with a capable wingman to deploy in battle. They therefore represent a step change in capability as a delivery platform for weapons and, at some $3 million apiece, a further route to achieving 'affordable mass'. There is unsurprisingly a flip side to this narrative. The US Navy has been successful in deploying high energy lasers, optical dazzlers and other close defence mechanisms against a portfolio of UAVs and out-of-the-loop surface craft. Other weapon systems will shortly employ short bursts of high-powered microwaves to disable swarm and other drone threats.

Uncrewed maritime surface vehicles are also upending *naval* balances of power given their ability to overcome currently available countermeasures. Ukraine's radio-controlled bomb boats have proved a sufficiently potent threat to cause Russian command to revise its force posture and deployment in the Black Sea. While developments in cheap, flexible sea-borne drones had already reshaped practice by the summer of 2023, much of this effect arises from the *threat* of the means rather than incidences of successful deployment and, in the case of Ukraine, the confinement of the Russian Navy to the assumed safety of its ports. Such threat, moreover, is magnified in confined harbour spaces and in low-readiness scenarios. At the very least, the deployment of UAVs must flex norms to prioritise new and expensive layered defence and the allocation of defence budgets to develop further countermeasures.[18]

Battlefield Empirics of Novel Systems

In considering procurement norms, disruptive technologies enable new models of engagement but also dislocate existing practices and focus

---

[17] Eric Lipton, 'A.I. Brings the Robot Wingman to Aerial Combat', *The New York Times*, 27 August 2023.

[18] Roland Oliphant, 'How Ukraine's drone Navy is menacing Russia's Superior Black Sea Forces', *Telegraph*, 26 November 2022.

planners on the value and fit of existing arsenals. After all, some newly procured equipment will be additive to those legacy means already in service but generally it should be assumed that deploying the latest hardware will quicken the obsolescence of long-dated kit and so require the development of new practices. This issue is relevant across domains as, from the perspective of forces' current portfolio of means, wide scope exists for new kit to disrupt, from synthetic-aperture radar satellites that can see through clouds to solid state vector sensors replacing hydrophones in the hunt for submarines. Planners (and norms) have always factored for threats to upend current practices, but it is the *breadth* of opportunity across procurement that today appears noteworthy. The norm for planners, after all, remains their duty to judge the importance and longevity of new assets to future battlefield practices but to do this recognising, of course, that a verso always exists; sensors require satellite bandwidth, sonar acoustics rely on vulnerable fibre-optic cabling and, ultimately, seeing is not the same as understanding.

While new technology can be a game changer, this may particularly be the case for smaller and non-peer parties. Well directed investment in these same innovative weapon systems might quite quickly overcome (or at least offset) the advantages of the well-resourced opponent. The planner's conundrum is that less resourced actors can now identify asymmetric advantages and exploit weaknesses through innovative and often low-cost means. Indeed, much Russian technology (here, its T-72 B3 tank platform, its Kalibr ship-borne cruise missile, its Iskander-M rocket and SU-35S aircraft) has often found itself blunted. And while such outcomes should not yet signpost new battlefield norms, the point also extends from the tactical to the strategic. After all, Russia's notion of 'cross-domain coercion', the blending of force with diplomacy, cyber and propaganda to achieve political aims, may resemble such a norm but, as a strategy, it has plainly not worked.

Although this analysis deliberately confines itself to exploring concepts rather than specific events that may or may not be relevant in a few years' time, a further procurement example is useful in considering how the acquisition of novel systems might move norms. Ukraine's deployment of Turkish Bayraktar TB2 drones stole early headlines and suggested a fundamental shift in battlefield risk and reward.[19] Small enough to be

---

[19] Stephen Witt, 'The Turkish Drone that Changed the Nature of Warfare', *New Yorker*, 9 May 2022, https://www.newyorker.com/magazine/2022/05/16/the-turkish-drone-that-changed-the-nature-of-warfare.

transported on a flatbed, the lethal drone has reportedly undertaken more than 1,000 strikes by the start of 2023 in conflicts from North Africa to the Caucasus. The bombs it carries can adjust trajectory mid-flight to deliver a payload that targets as precise as a trench. Capable of staying aloft for nearly 30 hours in its current iteration, the craft can fly to a height of 30,000 feet to conduct intelligence, surveillance and, using an on-board laser designator, to mark and engage targets with its four guided micro-missiles. Its range and payload may be considerably less than the US-made Reaper system, but this is to ignore its unique and disruptive advantage in its $5 million price tag. The American system costs upwards of $32 million for each platform. A key facility has been the TB2's ability to take out enemy anti-aircraft assets. It also has adjunct benefits from acting as a force multiplier (in particular its ability to optimise the delivery of artillery), an enabler of better decision-making and providing 'an unblinking eye' for parties' intelligence efforts. A further norm is that it has also displaced the need for vulnerable forward observation officers and, given its high-altitude loitering abilities, has proved itself a useful attritional asset in supporting land fires. These capabilities together suggest material change in means. A vehicle for other smart munitions, the drone has been an invaluable tool for opportunistic attack, either against enemy logistics or taking on high-value targets which were previously obscured by ground clutter. In the case of the Bayraktar, the system has also provided Turkey with political clout and an unforeseen strategic hand.[20]

The phenomenon is, of course, much broader than a supply of Turkish drones. The tasks undertaken by unmanned aerial vehicles are but a further step along a general continuum towards remote engagement but, when recently coupled with explosively formed penetrators and off-the-shelf technology (mapping, communications, networks, edge processing), they present as a properly new battlefield asset able to upend current doctrine. Turkey, moreover, has been willing to sell its technology to foreign parties to the alarm of Western powers. Countermeasures, however, have been quick to arrive and it may be that Ukraine represents a high-water mark in the capability's current guise and uncontested deployment, the more so given these systems' complicated supply lines and difficult procurement.

---

[20] Hambling, David, 'New Turkish Bayraktar drones still seem to be reaching Ukraine', *Forbes*, 10 May 2022, https://www.forbes.com/sites/davidhambling/2022/05/10/new-turkish-bayraktar-drones-still-seem-to-be-reaching-ukraine/?sh=7c85b64a685b.

The Primacy of Integration in the Deployment of Novel Systems

An enduring convention is that battlefield outcomes are very rarely dependent upon hardware procurement in isolation. Today's norm is that there is less requirement simply to kill the greatest number of one's adversary's population. Instead, victory is much more of a managed process that is shaped by getting the right narrative accepted by key audiences and maximising available assets by ensuring appropriate integration and to achieve this across existing means. On the battlefield, victory is achieved only by breaking the adversaries' combined will and, in operations, by being more resilient and more durable than one's enemies. These are not uniquely procurement issues as this analysis demonstrates the degree of second-order factors in war's prosecution. Procurement may define battlecraft and war's character but not its nature nor the list of long-held norms that govern war and warfighting.

Integration is the accompanying key to ensuring that novel systems are additive in legacy force design. This is to recognise the multiple sources of inertia that discourage the bedding in of new equipment, whether from poor motivation or leadership, user apprehension or poor understanding of the underlying means, risk aversion by commanders and operators alike, or because of cultural or other behavioural factors. Integration is also compromised by users' fear of failure, criticism, and career impact when results are not as expected. All of these traits are natural characteristics of service life but, as a norm, they have also combined to work against embracing change. It is usually easier to do less. After twenty years of service, the service personnel responsible for commissioning new kit may be better conditioned to solve individual issues rather than deliver new practices and to do this across departments.

Integration, moreover, is often ambitious, requiring bold bets in face of uncertain outcomes. The norm is therefore that integration of novel means requires a willingness to persevere despite setbacks, self-doubt and, germane to military settings, a likely loss of control. Without it, weapon platforms remain unoptimised, capabilities may be squandered and opportunity sets are lost. Technologies, moreover, do not necessarily all become relevant capabilities[21] but their best chance for this comes if their

---

[21] Breaking Defense Staff, 'Vehicle platform integration: Where technologies become capabilities', *Breaking Defence AUSA*, 14 October 2029, https://breakingdefense.com/2019/10/vehicle-platform-integration-where-technologies-become-capabilities/.

integration is driven by well-understood, well-led programmes of change that are collaborative, cross-domain and undertaken across the whole force structure. Integration and the embedding of new practices are, after all, fundamentally an exercise in behavioural science. Empirically, however, integration programmes are too often incremental. Plans of action are overtaken by other day-to-day priorities taking precedence, by a technology being introduced too early in its life cycle or, more often, by initial instances of disappointment then snowballing into general disenchantment, all of which create a brake to norm change. A verso, however, is provided again by the experience of Ukraine and its forces' clear trial of new means in the face of existential threat. Indeed, Ukraine's pace of innovation and the integration of new practices in an active war zone suggests, of course, that norms are likely to flex quicker and more profoundly in battle with technology and practice then spilling over to neighbouring parties. Forces in the throes of war become the catalyst for those adjacent parties to embrace change; in which case (the case, indeed, of previous generations), it is geo-politics that will again dictate the pace and degree of integration over the course of this primer's timeline, with the peril of war cancelling those day-to-day frictions set out above.

Norms and the 'Low Tech' Solution

Before closing out this section, consideration should return to today's asymmetry in available means. While the procurement challenges facing the West are expensive, uncertain, and long-dated, capabilities now available to non-state and non-war competitors can be inexpensive, irregular and innovative. An emerging norm therefore suggests an almost inverse relationship between complexity in new weapon systems and the 'low-tech', 'good enough' characteristics from off-the-shelf capabilities that can be pitched against them. Examples include spoofing, jamming, signal fratricide and denying one's adversary the means of parallelism. This gives rise to a further adapting norm whereby parties can increasingly compete *around* rather than against novel systems being integrated into the West's otherwise legacy force design. This also agitates for material change in practices. The norm, after all, remains that Western procurement relies on stringent bidding practices, open manufacturer competition and very frequent political interference. Its processes are byzantine, inflexible, and increasingly unsuited to the fast-paced flexibility required from today's force acquisition environment. Those practices should therefore change over the timeline of this primer, accelerated by the emerging gulf between institutionally sourced weaponry and commercially available technology that can be readily repurposed for the battlefield, ubiquitous and cheap to execute.

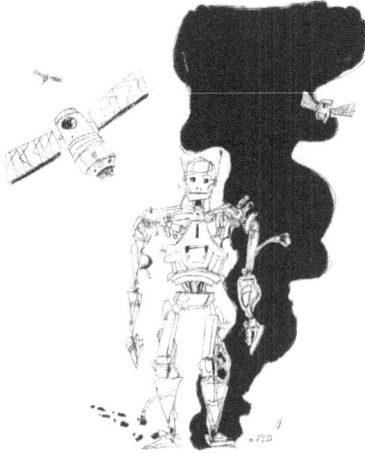

6
Autonomy and Thresholds of Supervision in Lethal Targeting

Battlefield deployment of artificial intelligence (AI), machine learning and weapons without human supervision is a near universal theme in commentators' assessment of future military structures and operation. Indeed, even evidence coming out of the Ukraine conflict suggests a 'slow but eventual, and seemingly unavoidable, evolution from military technology with the human fighter as the key performance metric toward one in which unmanned and autonomous systems will take on greater responsibilities'.[1] Furthermore, much *quicker* cycles of deployment, experimentation and evaluation all suggest acceleration in the pace where such systems will become, first, combat partners and then part of combatants' first line of attack.

Proponents of this narrative face remarkably little contradiction. After all, supported by dual-use technology developments, a portfolio of capabilities already underpins much of what is posited for battlecraft over the coming two decades. The norm change here is twofold. First, capabilities readily available from a myriad of sources, from retail communications, driverless car manufacturers, aviation as well as thriving data service industries, continue to be repurposed for military use. As this phenomenon

---

[1]   Samuel Bendett, 'To robot or not to robot? Past analysis of Russian military robotics and today's war in Ukraine', *War on the Rocks*, 30 June 2022, https://warontherocks.com/2022/06/ to-robot-or-not-to-robot-past-analysis-of-russian-military-robotics-and-todays-war-in-ukraine.

gains traction, the enduring norm whereby technology increasingly becomes a leveller to those waging war may be reinforced. In order to develop relevant capabilities, parties will need neither an institutional base nor state-sized budgets to acquire near-peer and peer means to wage conflict. The verso then becomes that parties must factor for the wide and efficient adoption of these new methods.

While most of this chapter deliberately considers empiric *challenges* to less human-centric and more autonomous processes on the battlefield, there is obvious advantage to be gained through machine-enabled decision-making. Indeed, even in Russian military parlance, autonomous capabilities are already lumped together in the catchall phrase of 'military robotics'.[2] For this to be properly disruptive, however, seismic developments are still required across a wide array of required componentry. While the explosion in data points that must underpin these capabilities have already been discussed,[3] the purpose of this chapter is instead to judge how norms might flex over the next 15 years as this process evolves, as less supervised systems are infused into military processes and frictions arise.

In this vein, there are obviously *degrees* of supervision. At one edge, full autonomy heralds the replacement of human supervision by machine oversight learning and by machine decision-making across a wide slice of battlefield processes. Indeed, at this end of the continuum, the decision to kill is being delegated to lines of code with algorithmic methods taking over the prioritisation of assignments, the execution of tasks, and the management and delivery of effect. A second theme from commentators then concerns robotics. These may either be human-in-the-loop/human-on-the-loop where meaningful human control (MHC) of such systems is always present or, again, with the machine operating autonomously and without supervision. The set of tasks envisaged is similarly wide, from logistical support roles to colleague machines that provide weapon platforms and where command and decision-making toggle between human soldier and that platform. The discontinuity here, however, is the ease with which control can be lessened across a portfolio of battlefield tasks, an example then being the code-dependent identification and prosecution of targets.

The norm here clearly remains for humans to manage all elements of battlefield decision-making. Indeed, arguments in favour of retaining

---

[2] For example, Russian unmanned initiatives include the *Marker* unmanned ground vehicle, the *Okhotnik* unmanned aerial combat vehicle, and *Vityaz* deep-diving autonomous underwater vehicle.

[3] In particular, see Chapter 2 (*Information and the New Importance of Data*).

MHC within AI-enabled weapon systems properly focus on safety, accountability, responsibility and dignity.[4] There is a widely held moral 'yuck' factor in delegating the decision to kill to an algorithm. In drone operations, passing norms are based upon at least three elements where the human manages the weapon system: first, defining the bounded area where the drone should search for targets; second, when confirming the target; and third, when authorising the intended plan of attack. Reducing human supervision in these processes requires substantial introduction of technology and wholesale changes to practices. It also complicates profoundly the role of the soldier in delivering military effect on the battlefield.

This is not surprising given that it is the very speed and complexity of data processing that are used to justify the role of artificial intelligence in processes precisely because humans can no longer match the speed or breadth of decision-making. The challenge remains, however, on how to retain human cognition and judgement in suddenly autonomous processes to ensure safety, accuracy and accountability. The expanded use of uncrewed assets in Ukraine will, after all, undoubtedly prompt changes in militaries' broadest ecosystems compared to how they looked immediately prior to Russia's invasion and, while limited to war's character (and not more), it is these developments that refocus attention on the ever-widening disparity between doctrines for war, training for war and the undertaking of war.

Human-out-of-the-loop platforms, after all, must be capable of selecting targets and delivering lethality with limited (or perhaps no) human interaction. Such weapon technology may still be in its infancy but both semi-autonomous and other similar precursor systems are already in service. Were these platforms to gain traction, they would certainly posit a new norm in warfare. There are, moreover, various drivers (removing humans from frontline danger; better capabilities in remote, hazardous, and testing environments; cost considerations; as well as a pathway to achieving force multiplication) to any move away from merely automatic weapons to fully autonomous weapons which are able to engage a target based solely upon algorithm-based decision-making. Indeed, while the norm remains that such evolution requires a material step change in both hardware and software, once deployed such machines would constitute a

---

4 Jovana Davidovic, 'What's Wrong with Wanting a "Human in the Loop"?', *War on the Rocks*, 23 June 2022, https://warontherocks.com/2022/06/whats-wrong-with-wanting-a-human-in-the-loop.

significant change in how humans wage war where either all or part of that system has independent agency and where the machine then has the ability to sense and act unilaterally depending on how it is programmed.

The continuum between automatic to autonomous weapons evidences that supervision thresholds are never binary and this must be reflected in how norms are flexed to accommodate increasingly remote practices in warfare. An emerging phenomenon, after all, is likely to be the deployment of *hybrid* weaponry whereby humans team with machines, retaining supervision until such time as the human is incapacitated or otherwise becomes an impediment to the team's performance. This is a difficult balance to achieve, not least given humans' poor performance at intervening in periods of stress or limited information. Indeed, norms should reflect that toggling control between human and weapon system is enduringly challenging. Those same technical difficulties (the management, for instance, of decision inputs that must empirically be based upon partial data, conflicting and duplicatory data, on obsolete data) may yet combine to prevent an effective self-learning weapon category from being legally deployed. Relevant challenges also include the setting of weapon values and goals, the anchoring of such weapons' internal representations as well as management of their utility functions, learning functions and other key operational routines.

Norms also need to reflect that it is enduringly problematic to code for ambiguity. It is similarly difficult for algorithms to factor for situational awareness or context. While the recent development pace of these technologies may appear extraordinary, fundamental fault lines endure complicating on how norms may move, not least the interdependent and highly coupled nature of the routines necessary for any dilution of the human's role in engagement sequences.[5] Nevertheless, the verso here remains that replacing human supervision in lethal engagement with algorithmic decision-making will render several well-tried concepts that have long comprised battlecraft no longer fit for purpose. It is this trying balance that will characterise norms in this fast-moving area over the period of consideration for this primer.

---

[5] Patrick Walker, 'Challenges to the deployment of autonomous weapons system', Buckingham University Humanities Research Institute, August 2019, https://www.academia.edu/83823709/Challenges_to_the_deployment_of_autonomous_weapons and Patrick Walker, 'War without Oversight: Challenges to the Deployment of Autonomous Weapons', August 2019, https://papers.ssrn.com/sol3/papers.cfm?abstract_id=3757516.

Norms around Remote Engagement

In understanding norm change in this space, it is useful to consider further the incentives that exist to move parties towards adopting remote warfare in their battlefield operations. Uncrewed systems promise force multiplication, precise warfare and, given politicians' aversion to casualties, a means to lessen risk while leveraging technical superiority. Proponents also suggest that unsupervised weapon systems which are constrained by rigorous programming might offer better *ethical* performance in times of stress than their crewed alternative. The development also promises to mitigate manning shortfalls. It posits a reduction in the burdensome costs of human resources in their several guises, be they soldiers' benefit bill, the inconveniences of sleep patterns, or the fiscal and tactical costs of casualty evacuations and care. Norms should also recognise that the trend continues to be accelerated by the private sector and, for decision-makers, is also conditioned by that same 'revolution in expectation' ('I have seen it on a screen and therefore it exists') that highlights the *art* of the possible rather than the realities of what is technically and practically feasible.

Several drivers exist, therefore, that promise improvement in one or other battlefield processes whether that be better operational performance, broader combat options or advantageous expansion of the battlefield into new arenas. In terms of empirics, however, neither the efficacy nor subsequent integration of these technologies is clear cut. The long list of capabilities that must be in place before unsupervised machines can demonstrate predictable operation includes several enduringly difficult processes such as extrapolative reasoning and learning informed from the weapon's sensed data. Autonomy, after all, is fundamentally data analytics at scale and therefore requires an unbroken means to label and classify data arriving from its sensors before such inputs can become the basis for meaningful decisions absent from human intervention. The seamless operation of these processes is, after all, pivotal if unsupervised lethal engagements are ever to be militarily relevant and, moreover, ethically acceptable.[6]

Operational considerations arising from the deployment of uncrewed aerial vehicles (UAVs) are considered in the previous chapter, but it is technology's *autonomous* promise that will likely have more impact on

---

[6] Patrick Walker, 'Leadership Challenges from the Deployment of Autonomous Weapons Systems; How Erosion of Human Supervision over Lethal Engagement Will Impact How Commanders Exercise Leadership', *RUSI Journal*, July 2021, https://www.tandfonline.com/doi/full/10.1080/03071847.2021.1915702.

battlefield norms. While the platforms' underlying componentry already promises users new flexibility in the method, degree and pace of operation, whether that be the remote gathering of information or in remote means of attack, it is the independent and unsupervised use of these uncrewed platforms that promises to upend commanders' modus operandi. For this to occur, however, several sources of friction that remain along this pathway must be overcome. First, doctrine and tactics must be developed to ensure that current practices are amended in a manner that still realises these platforms' potential. As is so often the case, a further verso here is that the scale at which new means are anticipated to change the future battlefield is very likely inflated. Second, forecasts routinely overlook the costs and complexities of their introduction. This list of challenges looks daunting and includes the complexities of configuring the technology, costs arising from maintenance and logistics and the in-situ competencies required to deploy these fast-changing new assets. Specific difficulties related to UAVs occur around the launch and recovery of these platforms, power issues, the need to manage communications and master all-weather skills as well as the assets' ability to survive when close to threat.

Deployment of uncrewed assets, whether in or out of the loop, has other ramifications. In articulating norms, debate will remain around *which* capabilities should be delegated to uncrewed systems. Here, the Ukraine conflict has demonstrated that new technology soon becomes a priority for the enemy's targeting. UAV assets will likely be among the first capabilities that a future force will lose even before a battle begins. This too has norm consequences for decision-makers considering their irrational deployment in order to prevent degradation, increasing the risk of escalation while reducing future flexibility and complicating planning given the procurement lead-times of these assets. A behavioural concern then arises from commanders' increasing reliance on uncrewed technology. Old skills can be quickly lost, and experience in basic, foundational practices vanishes. Battlecraft is then disproportionately impacted should new technology fail or be compromised. Diminished situational awareness, for example, can rapidly impact the commander's sense of vulnerability, lowering the self-defence threshold in ambiguous scenarios or increasing incentives for pre-emption, the notion of use-it-or-lose-it.

Challenges to the Deployment of Autonomous Systems

Particular difficulties obviously arise with the deployment of unsupervised, autonomous assets. Taking away the human operator requires an alternative decision-making solution to be substituted in place of the human and several parts of this technical puzzle remain enduringly out of reach. Missing pieces include reliable target recognition that is resilient to feint and adversarial meddling, the weapon's ability to manage ambiguity, achieving (and then maintaining) appropriate situational awareness and having the machine properly understand context. There are also behavioural challenges. The complex, probabilistic underpinnings of the autonomous weapon complicate how and when commanders use these independent assets. As currently predicated, machine-learning routines that will underpin prediction and optimisation routines appear far too susceptible to adversarial countermeasures and spoofing. Weapons that have been trained, say, to identify a tank through its box-like shape, traditional turret configuration, movement pattern and radio frequency profile can be fooled into ignoring threats where signatures do not exactly match. An unexpected colourway, an adjusted parameter, a deliberately changed property may lead to that that enemy asset passing unengaged. This is obviously unacceptable.

As currently envisioned, the data components of autonomous routines fail to backfill for incomplete or fractured inputs. This has operational but also legal consequences. Operationally, machine learning's current technical spine inadvertently leads to the suppressing of doubt in its data sets as well as inappropriate inference of causes and the narrowing of choices that the lethal machine will consider. Indeed, the position of this primer is that too many foundational functions remain outstanding for human agency to be delegated to algorithmic machines. These include appropriate mechanisms for the weapon to forget data, proven routines for the weapon to undertake unsupervised verification and validation, the configuration challenges around a weapon system that is expected to learn from its surroundings as well, generally, as means to anchor and adjust the machine incrementally for recently sensed data.

Even if it were possible to create relevant training datasets for these systems, changes in human behaviour will likely make any initial training data immediately invalid. This is particularly true in conflict situations where combatants very quickly change behaviours in the heat of battle, such changes not being reflected in these training sets. Efforts, moreover, will always be made by combatants to evade detection and sow confusion, further compromising these platforms' ability to find patterns in their observed surroundings. In considering norms, moreover, technology empirically confers only fleeting advantage to the adopter with adversaries either copying processes or devising workarounds.

A further emerging norm around autonomy in systems must be that irregular and non-state actors will likely have less compunction about deploying similarly independent weapon technology, the more so if those parties judge it can deliver some level of accretive effect regardless of legal shortfalls. In the case of irregular forces, the difference between an autonomous weapon and a drone is only four exploding bolts as the limitations of the Law of Armed Conflict, ethics and precedent will likely be trumped by non-state priorities of expediency, belief and culture.

Human-Machine Teaming

Just as new technologies have long been a driver of military advantage, the consensus is clearly that robotics and artificial intelligence offer particular potential for another inflexion point in delivering military transformation. Notwithstanding skills shortage in the vertical, the accepted narrative is that parties able to acquire the best AI and robotic systems should gain significant advantage over those with less capable technology.

Teaming solutions, the combination of humans and colleague machines to leverage the capabilities of trained soldiers with the increasing utility of machines, are likely to become the technology's first deployed manifestation over the timeline of this primer. The vision for these teams is that collaborations must perform at least as well as soldier-only groupings but with greater adaptability, better protection and options for the human in the grouping. It is also for configurations that are easier to put together under engagement and all then empowered by informed and quicker decision-making. In this vein, teaming promises to reduce both the number of humans involved and the risk to which those humans are exposed while achieving force multiplication and broader tasking at reduced cost.

There are, unsurprisingly, several challenges to the deployment of human-agent teams before norms are likely to be impacted. A reliable and easy-to-execute basis of understanding between soldier and the machine agent remains outstanding. While a basis for this collaboration is that colleague machines leverage the valuable experience of the human soldier, the aim of hybrid teaming must be that machines will in time operate autonomously at least during phases of their combined operation with the human counterpart. For this to occur, the machine must be able to factor for human tasks and workloads, task complexity, and the planning and execution of subsequent phases across operations.

Robust methods must similarly be in place to predict performance across these combinations before war planners and commanders can delegate tasks to these novel groupings. In human-only teams, colleagues clearly perform complementary, largely non-redundant functions. In human-machine teams, it will be necessary to share knowledge, awareness and learning in lockstep with the unfolding of what are complicated, nuanced tasks. Indeed, all of the mechanics and constraints of task autonomy, even if this is somehow to be bounded and controlled by the platform's initial configuration, will remain a prerequisite for norm change if properly disruptive teaming is to be deployed.

Deployment challenges to human-machine teaming are also likely to be *institutional* with consequent restriction upon the speed and degree of norm change from the technology. Tactics and doctrine must adjust to manage the joint working that will need to underpin these new configurations including protocols to facilitate the toggling of command and control between human and machine while recognising humans' generally poor cognitive performance when confronted with new tasks in times of stress and partial information. An adjunct concern remains the lack of consensus on the certification, verification and validation of self-learning weapon systems, the more so in adversarial environments where humans and artificial intelligence will be expected to cooperate.

Again, the degree of change to norms will be governed (and delayed) by what it will mean *empirically* for humans to partner with technology on the future battlefield, by issues of agency and authorisation, and by well-understood protocols for these delegation mechanisms. Tabling these teaming protocols will be no easy matter as they must factor for difficult, subjective issues such as intent, relative capability, context (that is dependent on the particular and fast-changing environment that the team is operating), as well as task definition and learning in settings without clear abstraction or boundaries. Overlaying legal considerations (whether that be laws of armed combat or particular rules of engagement that are relevant to the theatre of combat) and doing this, in the case of the weapon platform, through the medium of code alone, would seem to be an overly complex set of tasks.

In judging norm movement in teaming, it will also be necessary to account for the phenomenon's *social* components. This is complicated on several levels, not least the human operator's propensity to ascribe social intentions to machines where none exist. Other factors have ramifications. Can a machine be heroic? Can it engender respect? How does machine operation dovetail into basic and long-held tenets of leadership? Indeed, norms here must reflect that teaming cannot add value to all relevant tasks and must of course reflect what is a broad continuum between, at one end, an otherwise dumb weapon and, at the other, the sophistication, preparation and experience of the trained soldier. Two observations arise. First, mechanisms that foster trust and demonstrate assurance will be key to teaming's integration in otherwise legacy assets and means. Second, the success of teaming over this primer's timeline will rely more upon entirely new models of deployment rather than mere overlaying of new technical capabilities upon existing modus operandi. Simple replacement of digital transformation on top of current practices will neither work nor change current norms.

Similar to integrating aviation into navies' fleet operations prior to World War II, deployment of human machine teaming will require a period of intense experimentation. Norm change will lag correspondingly. Trialling and making mistakes, after all, is never an easy space in which to assess changes to behaviours until clarity on outcomes can be achieved. Nor should such teaming be regarded as a standalone capability given that its successful integration will be dependent upon other assets that constitute parties' battlecraft assets at any point in time. Indeed, benefits must outweigh costs throughout teaming's introduction.

7
Battlespace Fighting
Changes to Operations in Rear and Close Quarters

A recurring theme for the primer's initial commentators was broad agreement around recent material changes to sensor densities that actors can now deploy across the battlefield to observe opponents and their behaviours. This may seem a strange topic to open a chapter on battlespace fighting but it illustrates the degree and speed of change upending practices. A clear new norm is therefore that forces should expect to operate effectively detectable on the battlefield, local intelligence on their positions and intentions now buttressed by citizens with their phone cameras, by real-time open-source intelligence and now by ubiquitous sensor coverage noting and relaying troop configurations from every angle. The position and movement of assets are now continuously surveilled by satellite, ground-moving-target-indication radar, uncrewed aerial vehicles and sensitive infrared platforms regardless of terrain and, largely, of countermeasure. In conjunction with later commentary considering developments across the battlefield's electro-magnetic spectrum, this section now reviews norm change arising from in-theatre advances in the practice and technology of fighting.

New Assets and Norms

Parties ignoring changes in battlespace risk fighting their next war with their last war's capabilities. Nevertheless, while debate continues to be dominated by numbers and platforms, by intent and available means of combat, the prevailing norm is that bases of operations remain little changed. Land mass has not shrunk and forces are still able to move around at the battlefield's many seams, the enduring notions of a 'hay-in-a-haystack' and 'hiding in plain sight'. Adversarial activities still hamper the observer's ability to abstract and the many inputs that now constitute a commander's decision-making remain just as compromised by ambiguity, temporality and biases. While the coverage of passive sensors may have increased substantially, appropriate verification of intelligence sources as well as the situational awareness that is derived from these sources is still required to inform decision-making, whether this be undertaken from stand-in sensors or by the eye of the soldier. Layers of sensor technology may provide interesting detail upon which to hatch plans, but recent conflicts prove again and again that seeing does not equate to understanding. The notion that all shooters will receive all available information all of the time remains enduringly challenging.

While technology offering such awareness may suggest delineated, well-understood battlespace, this remains unproven and currently falls considerably short of a discontinuity with which to refine current norms.[1] Relying, after all, on software for such key fighting capabilities risks much should those applications fail to perform as predicted.[2] Decision-makers may become over-reliant upon third party aids and those aids may be easy to compromise. Electro-magnetic leakage, moreover, is a dangerous product of militaries' growing dependence on data, providing new target sets for enemy assets to identify and engage.

As demonstrated throughout the primer, however, a key driver for norm change remains developments in battlefield equipment. Modern combat is still based upon the taking and holding of ground and today's strategic calculus around armour, infantry and precision fires would be immediately recognised by earlier generations. Nevertheless, this masks quite fundamental changes in battlespace, not least from new means of gaining advantage springing from unexpected quarters, from retail and commercial applications of technology and practices that have no direct military genesis. Communications, navigation, data dissemination and manipulation are stock capabilities that may be taken for granted but all of which are now foundational for military processes. Indeed, universal

---

[1] See, for instance, https://www.palantir.com/platforms.
[2] Economist editorial, 'The Fog of War May Confound Weapons That Think for Themselves', *Economist*, 27 May 2021, https://www.economist.com/science-and-technology/2021/05/26/the-fog-of-war-may-confound-weapons-that-think-for-themselves.

access to technologies such as online navigation, surveillance and the management and optimisation of assets should now be regarded as a further new given in combat operations.

The norm is therefore that these conditions can now be achieved regardless of an adversary's status as either near-peer or non-peer; all combatants, all of one's enemies now possess seamless mapping, surveillance and organisational tools. Workflows can be optimised and priorities tuned to suit available resources in a manner previously unavailable to planners and commanders. A flip side is that governments will be reluctant to degrade or halt such information systems in times of stress, afraid of hampering their own operations but also of creating alarm in their own populace. These software and phone applications are, after all, ubiquitous and as foundational to the lives of inhabitants in a combat zone as to those trying to take that zone by force. This is not an obvious development. While combatants can leverage such tools regardless of battlefield quarter (whether that be close, near or far), the same works for both the offence and the defence. This too has consequences. While images today of flattened residential and pummelled infrastructure may present as wars of the past, the assets wreaking this damage must themselves disperse to avoid remote engagement and attrition. Indeed, the British Army's Strike doctrine calls for a 2,000 kilometre buffer before joining the close battle.[3] That is an extraordinarily long distance to be coordinating assets and managing risk.

A developing norm is therefore that previously safe rear areas have vanished and that it is no longer really possible to quarantine areas of the battlefield. The same holds true of airspace: Ukraine has demonstrated mere possession of aerial assets is increasingly removed from achieving air superiority. The follow-on here is a considerable expansion in the size of possible operational areas requiring commanders to bring together thinly dispersed forces (particularly air assets) at a particular time and at a particular point in order to create effect while still preserving assets in the run-up to action. This then complicates reliable command and control infrastructure which must now operate across this much larger, exposed and observed footprint. Conversely, distance increases both its fragility and the priority attached to such assets that must control the fight regardless of electronic attack (and do so in this expanded space). The long-dated norm

---

[3] Wavell Room Publications, 'Big Toys: British, Ground-Launched, Long-Range, Precision Strikes', 19 September 2019, https://wavellroom.com/2019/09/19/big-toys-the-requirement-for-a-british-ground-launched-long-range-precision-strike-capability/.

around commanders' protective bubbles-of-concentration is ever more relevant given the vulnerability to logistics and digital assets now presents.

An adjunct norm in this vertical arises from actors' opportunity to leverage velocity and adaption. The norm here is that improvements in weapon performance will continue to increase the lethality and depth of engagement, a particular factor at the outbreak of hostilities for which commanders must factor in opening their campaigns.[4] An adjacent norm, however, is that predicting the trajectory of battlespace fighting will remain a frustrating and hazardous exercise. Basing plans on recent conflicts is to risk shaping responses and priorities on assumptions that are already out of date, whether this is the framing of expectations around a particular technology that does not deliver (or, as likely, is quickly superseded) or upon sets of circumstances that prove irrelevant in light of an adversaries' ability to surprise.[5] The developing norm around battlespace should not, therefore, be for any change in strategic timeframe; after all, it still requires a decade or so to create a new division given the friction of training, doctrine, establishing a relevant experience base for any such structure and then satisfactorily equipping it for its new tasks. What remains immutable, moreover, is that forces must first identify an adversary's hostile actions, coalesce around a response to those activities, then navigate themselves towards that first engagement in a timely and prepared manner while marshalling sufficient political will to engage in and sustain prolonged military operations.

Russia's offensive in Ukraine confirms the challenges of making sweeping conclusions on warfare's future based on a single episode. Ukraine is but one conflict and will certainly differ from the next flashpoint. The norm here remains that any observed set of conditions is unlikely to be either exactly relevant or precisely replicable to a party's next conflict. Playbooks differ for each subsequent outing as tactics and technologies that are effective in one setting may be inappropriate for the next. Nevertheless, fighting the next war with means that are shaped by earlier campaigns might be an uncomfortable axiom, but is a norm in itself. Commanders rarely go to war with what they consider to be an ideal set of assets but this can be mitigated by adaption, by reformist command and by understanding

---

[4]   Dan Baer, 'Six reflections of the first day of Russia's war in Ukraine', Carnegie Endowment for International Peace, 24 February 2022, https://carnegieendowment.org/2022/02/24/six-reflections-on-first-day-of-russia-s-war-in-ukraine-pub-86524.
[5]   Bonnie Berkowitz and Artur Galocha, 'Why the Russian military is bogged down by logistics in Ukraine, *Washington Post*, 30 March 2022, https://www.washingtonpost.com/world/2022/03/30/russia-military-logistics-supply-chain/.

the passing norms of warfare (or at least their drivers). Russia's early tank losses in Ukraine, for instance, were not the product of having to use its last war's kit. Instead, it happened because of poor deployment decisions, poor planning, inadequate combined arms support and mistaken expectation around the response that the invading force would encounter (both tactically but also logistically in terms of assets deployed) from its adversary. As many as half of Russia's lost tanks in 2021 and 2022 may first have been abandoned by their crews. Ascribing norm change is therefore challenging without clear abstraction between Ukraine's modern anti-tank guided missiles, inappropriate tactics, absent management, Russia's logistics gridlock, poor kit, corruption's effect on its supply chains et al.

It was also likely that it was Ukrainian artillery which wrought most damage to these assets. Indeed, hasty norm change should be dampened by exactly the long-held and repeating rhythm of design, redesign, adaption, and trialling that has long underpinned the development and deployment of military equipment; while tank capabilities have grown, so have threats facing and then, in turn, countermeasures to these threats. This explains the general longevity of weapon types and why it is difficult to call the end of a battlefield asset class. The survivability of today's tanks, moreover, remains considerably greater than that of other armoured vehicles and, without tanks, parties intent on large-scale ground war must rely on other fighting platforms to fill that same role.[6]

Rear and Deep Operations

An emerging norm is that it is now difficult for forces to rely on their rear areas as a place to reorganise, replenish and take stock. Previously safe zones located considerably behind front lines are no longer unseen or out of reach of adversarial action. Both parties' use of long-range munitions, both precision and unguided, against rear echelon targets in Ukraine has reinforced this norm, its consequences being to constrain logistics, to threaten infrastructure located away from any front as well as to destabilise local population concentrations. Being able to influence a party's 'deep' massively expands the size of the operational area that defending parties must consider in play. This in turn complicates communications, complicates replenishment as well

---

[6] Rob Lee, 'The Tank Is Not Obsolete, and Other Observations about the Future of Combat', *War on the Rocks*, 6 September 2022, https://warontherocks.com/2022/09/the-tank-is-not-obsolete-and-other-observations-about-the-future-of-combat.

as complicates the coordination of now dispersed assets in this expanded terrain. Similarly, activities in such previously safe haven areas now require their own quite separate logistics, footprints and digital signatures, each creating new levels of attack surface and vulnerability.

Even before Russia's invasion of Ukraine, developments in sensor and fires technology (as seen, for instance, in Nagorno-Karabakh, Syria and Libya) suggested that armoured manoeuvre in rear areas already presented considerable challenge. Parties were already able to identify and then engage targets well before traditional sighting of enemy assets had been recorded and sent up channels of command. Ukraine's invasion has therefore highlighted the need for a change in practice.[7] While several change agents in this matter have been strategic (the same litany of poor leadership, centralised command processes that were not fit for purpose exacerbated by an absence of non-commissioned officers to provide stability and ballast to Russian battlecraft), others have been tactical. These include withdrawal of sensitive equipment to avoid its capture by Ukrainian forces, inadequate capacity whereby Russia expected to hold ground with just four soldiers per thousand of Ukraine's population, insecure communications and failure to achieve air superiority, both sets of drivers coming together to aggravate weaknesses and compromise activity in those areas deep behind parties' front lines.

Lessons, moreover, arise for the UK's own current force posture and how it conducts operations in its rear. While not themselves norms, the examples provide useful pointers to inform future force adaption, procurement priorities and, in time, future responses to an adversary's playbook as it plays out. British self-propelled artillery, for instance, lacks both range and the number of platforms to influence conventional conflict with near-peer competitors. The UK's stock of multiple launch rocket systems cannot then be fielded either in time or in sufficient numbers to halt peer forces behind their lines. Other line-item challenges exist. The Tornado, Britain's premier ground-attack plane through the latter stages of the Cold War, has long been retired without replacement and the F-35C is unlikely to be bought in the numbers required to provide like-for-like capability on the future battlefield. Until Russia's invasion of Ukraine, Britain's tank fleet was set to shrink by a third, the equivalent of just two fighting regiments

[7] Seth Jones, 'Russia's ill-fated invasion of Ukraine: Lessons in Modern Warfare, CSIS, 1 June 2022, https://www.csis.org/analysis/russias-ill-fated-invasion-ukraine-lessons-modern-warfare.

plus training platforms and spares. While the Challenger may remain a useful fighting vehicle, at best it represents parity rather than overwhelming superiority against likely adversaries' equipment. These developments point to a further developing norm, that of the curiously sparse battlefield. As noted by the *Economist*, some 350,000 Russian troops were arrayed in Ukraine in the summer of 2023 and this along a front line that stretches for 1,200 kilometres. That is around one tenth of the average for the same area in the World War II. Battalions of a few hundred men now fill areas that would once have been occupied by brigades of a few thousand.[8]

Norms around the Urban Domain

This project's evidence base also noted several warfighting norms undergoing significant change, none more so that the priority it placed upon urban settings as the pivotal future battleground. This is unsurprising given the extraordinary expansion of the urban and built-up environment. Militaries, conversely, have been shrinking over the same period. The developing norm is that towns and small cities will no longer simply be subsumed as an army moves through. The difficulties, moreover, involved in fighting and deploying technology into towns and cities (and then bringing force to bear in those environments) will increasingly make urban settings the preferred battlespace for defenders.[9] In particular, insurgents can draw out conflict in the built environment.

Gone, therefore, is the notion of a quick war where manoeuvre equals 'smart' fighting equals limited and controllable battlespace. Instead, Ukraine (and Afghanistan and Iraq immediately before) should be suggesting that actors commit financially to their militaries as the tempo and duration of conflict becomes less and less manageable. The norm here remains that it is one's adversary that dictates what force size makes sense. In this vein, passing norms around urban war have various components. First, an enduring behaviour must be that urban mass discourages troop concentration and, in degrading communications and hierarchies, it blunts attack in size. Buildings provide cover, allowing defenders to prepare and then dominate wide areas of the built environment. It shields the defender's

---

[8] Economist Editorial, 'Ypres with AI', *Economist Special Report on Warfare After Ukraine*, 8 July 2023.

[9] Margarita Konaev and Kirsten Brathwaite, 'Russia's urban warfare predictably struggles: fighting in cities is hard for any military', *Foreign Policy*, 4 April 2022, https://foreignpolicy.com/2022/04/04/russia-ukraine-urban-warfare-kyiv-mariupol/.

strength and intended pattern of fighting. Urban mass grossly complicates the role of the attacker whereby every obstacle becomes a potential attack angle, to be investigated and neutralised before being considered safe. But defenders can move quickly to reoccupy that previously safe structure. Attackers' slow progress allows defenders time to fortify positions. For this reason, fighting in towns generally (and literally) levels the playing field, almost regardless of adversaries' relative size.

Sensors are also less effective in the complex-built environment, reducing an attacker's ability to be situationally aware. Visibility in the urban space is challenging, the more so once defenders have been able to fortify their domain. The reach of technology is diminished; even night-vision goggles are compromised as they rely on amplifying faint ambient light which is absent in subterranean environments. Infrastructure may be generally more durable (or, at least, fixable) but, as Ukraine has experienced, is not necessarily any more difficult to target. Indeed, a dreadful enduring norm remains that attrition of an adversary's civilian infrastructure remains a key strategy in modern conflict in order to terrorise civilian populations, precipitate unrest and the flight of refugees and, in so doing, pressure administrative organs in those territories under attack. Urban environments are also target-rich environments although it remains difficult to prevent defenders' re-infiltration of space that has been demolished and infrastructure that has been broken. Should that attacker be intent on occupying urban space, it must still commit to allocating mass for cordoning and screening operations.

It is the example of Mariupol's grim experience in 2022 that reinforces how little real change there has been in norms around this area. Indeed, battles in cities have been central to several recent conflicts including Shusha in Nagorno-Karabakh, Mosul in Iraq and Raqqa in Syria, all reinforcing the enduring norms around the tactics, importance and complexity of urban warfare. After all, over half of the world's inhabitants now live in towns and cities, a figure that is expected to rise to two thirds by 2050. In Taiwan, for instance, 80 per cent of its population is urban. Moreover, as cities have swollen, armies have shrunk, less able to swamp an urban sprawl or to surround them with multiple fronts. The consequence here (and an enduring norm) is the ongoing role of the micro-siege, quite often individual structures where a single building can consume an entire battalion over days of fighting. A second consequence, also not new, is that nine in every ten casualties in populated areas are likely to be civilian. That precision bombing (when it is available and when it achieves effect) can target specific buildings is less relevant when combatants can simply move next door such that bombing must follow defenders from house to house. And Ukraine's defence against invasion by its neighbour has reinforced that precision munitions are both expensive and in short supply.

A further feature and enduring norm of urban warfare is its underground environment. Many of the newer technologies listed in this primer simply do not function below the surface, especially satellite navigation and drone surveillance. Again, a norm remains that the general built environment disproportionately degrades the efficacy of the more technical fighting force. Adversaries can only see what they can physically see with large segments of enemy activity and assets remaining hidden and unknown. Barriers channel attackers. Streets, alleyways and doors will be blocked, turning urban areas into a fortress of walls concealing the defender's prepared traps. Here, Ukraine's Mariupol demonstrates that no amount of aerial intelligence can pierce defenders' concealment and underground movement, passing the initiative to defenders and, to a greater extent than in open countryside, to their timetable. The norm here is therefore that the urban terrain will continue to disadvantage the attacker in terms of intelligence, surveillance, reconnaissance and the ability to engage at distance. The attacker must continue to fight with limited cover and must still negotiate every building prepared as a fortified bunker.

A more important norm arising from the first 14 months of the Ukraine conflict is that urban areas are becoming de facto sanctuaries for actors' assets. As technology (such as multi spectral sensors) penetrates cloud or dark, unhindered passage for military forces moving in rear areas and between towns and cities is now no longer available. This is a discontinuity and one that reinforces the urban norms set out in this section. For both attacker *and* defender alike, adding to a city's existing political and economic value is increasingly a haven effect of being able to obfuscate and muddy detection and house military hardware in relative safety (relative to previous practice of corralling those assets in plain sight).

A follow-on consequence (and strengthening norm) is that each small town can readily become its own citadel, easy to defend with few assets yet ever more difficult to take. Able to act as a shelter for subsequent raids, urban areas can less and less be bypassed and become tactically more challenging for attacking forces. Indeed, rates of advance are one third to a half that which commanders might expect and plan in non-urban combat, with a recent US manual recommending that the offensive should outnumber those defending an urban setting by as much as 15:1. The norm-affecting verso here is that resupply, coordination and mobility for the urban defender are very challenging. Large numbers of defenders can themselves be pinned down in a small number of non-contiguous locations that can be picked off, besieged or left for later engagement. Similarly, multi-pronged advances can paralyse the defender's decision-making and flexibility.

Finally to this point (and itself an enduring norm) is that urban warfare will continue to complicate the relationship between civilian and combatant. Previously confined to an essentially passive role, city-dwellers in modern conflicts (in, for instance, Ukraine) are not only armed at war's outset by local governments but, in itself a discontinuity, residents today can become a technically enabled informal network of spotters and facilitators as well as being an armed adjunct to their regular army.

Norms in New Domains

Air superiority is generally assumed to have been achieved when one's enemy is prevented from flying while retaining the ability to fly oneself. This view remains central to the doctrine of Western air forces and has previously informed norms in this space.[10] Today, however, this involves a complicated set of equations with several inputs informing the duration, depth and state of that superiority. For instance, the early phases of the war in Ukraine failed to establish Russian air superiority, despite Russia's large and technologically sophisticated air force. This also suggests an emerging norm whereby denying air superiority becomes a smarter operational objective than trying to gain it outright, exploiting the potential of air denial and strategies designed instead to exercise limited control of airspace or temporary, localised air superiority.

In this vein, Russia's vaunted air assets have floundered for several reasons (poor maintenance and spares availability, poor logistics and refuelling) but also because of systems' systemic reliance upon data to deliver effect in difficult, adversarial environments.[11] The fog of war has again shown data to be enduringly patchy and ambiguous, degrading those systems' promised capabilities. It is the authors' contention that the

---

[10] Maximilian Bremer and Kelly Grieco, 'In denial about denial: Why Ukraine's air success should worry the West', *War on the Rocks*, 15 June 2022, https://warontherocks.com/2022/06/in-denial-about-denial-why-ukraines-air-success-should-worry-the-west.

[11] Steve Inskeep, 'A big mystery of the war in Ukraine is Russia's failure to gain control of the sky', *National Public Radio*, 11 May 2022, https://www.npr.org/2022/05/11/1098150747/a-big-mystery-of-the-war-in-ukraine-is-russias-failure-to-gain-control-of-the-sk?t=1654508553446.

effects of data fragility on combat operations remain under appreciated with ongoing consequences for war's forms and norms. It also presses for continued meaningful human oversight. Empirically, data friction actually requires that initiative be passed straight back to human defenders who are local to the unfolding action. It unravels processes and can soon reveal those frontline operators' lack of training. Without efficient data, machine processes fail and battlecraft collapses back to sticks and stones.

Ukraine's air denial practices have been informed in the 14 months to June 2023 by this new set of behaviours. They have instead relied upon nimble 'shoot and scoot' tactics to ensure their own service-to-air missile assets remain fleeting targets. Russia has then had to rely upon stand-off sensors to identify radar targets, lengthening the time required to engage such mobile systems. Flying low in order to evade radar detection has then exposed Russian aircraft to enemy anti-aircraft artillery, thousands of shoulder-fired, soldier-portable air defence systems and locally crafted air defence traps. It is these three tandem components, an amalgam of new hardware accompanied by tactics tailored to the new capabilities that these platforms now provide, that suggests a developing norm. Even if Ukraine cannot secure air superiority for itself, notes Bremer and Grieco, it has still been able to deny it to the Russians. Examples, moreover, drawn from Iran, Israel, ISIS in Syria, Nagorno-Karabakh highlight that 'the global spread of advanced, highly mobile long-range surface-to-air missiles, man-portable air defence systems, and loitering munitions along with continued advances in networked unmanned systems, dual-use robotics, sensors and advanced materials, place the capabilities needed to contest air control in more adversaries' hands'.[12]

This move will only accelerate as uncrewed, nearly autonomous systems are introduced, able to engage in swarming tactics that involve thousands of cheap small-sized drones. Indeed, old-fashioned air power has kept what is a surprisingly low profile in the Ukrainian conflict. This suggests an emerging norm. Aviation assets are fragile, expensive and difficult to replace. Any losses are disproportionately noticeable and easy grist to adversaries' narratives. They have been eclipsed instead by kamikaze drones, precision strikes and cruise missiles deployed by both sides. The stepped-up air activity has generally resulted in unexpected

---

[12] Maximilian Bremer and Kelly Grieco, 'In denial about denial: Why Ukraine's air success should worry the West', *War on the Rocks*, 15 June 2022, https://warontherocks.com/2022/06/in-denial-about-denial-why-ukraines-air-success-should-worry-the-west.

loss rates with close-air support aircraft faring particularly badly.[13] Flying so close to enemy assets in support of ground troops puts such aircraft in harm's way requiring enhanced protection both in terms of armour and decoy measures, degrading performance and likely limiting how such assets may be deployed on the future battlefield. While this might suggest the developing norm for air power, more peer-on-peer conflict (or air combat where one party materially outguns its adversary) might see air deployment play out quite differently.

Analysis of battlespace changes must also cover the emergence of near and deep space as a newly critical, contested, congested, and competed battlespace. While space has long been pivotal to terrestrial operations, the domain is certain to increase in importance over the timespan of this project, in part because the current legal framework provided by the 1967 Outer Space Treaty is no longer fit for purpose and currently allows actors considerable latitude in adversarial action. Interfering, for instance, with parties' positional, navigational and timing (PNT) assets would have significant consequences for the battlespace, the more so given the inadequate priority and budgets (and, in the case of NATO, legal and societal constraints) to training without space-based communications, space-based surveillance as well as control and command undertaken without a satellite component.

Rapid development in space technology, furthermore, reinforces this pending norm change. Expected assets include the deployment of nanosatellites, below-orbit PNT, rendezvous-proximity operations and soft-kill mechanisms of such space assets. Space developments also have a deterrence dimension given the plethora of inexpensive launch vehicles and the destabilising effects from asset and capability proliferation in the domain.[14]

Conflict in the Electro-Magnetic Spectrum

Notwithstanding the poor delineation between electronic warfare and cyber capabilities in public debate, breakneck developments continue in

---

[13] Economist editorial, 'Has the Ukraine War Killed Off the Ground-Attack Aircraft?', *Economist*, 1 November 2022, https://www.economist.com/the-economist-explains/2022/11/01/has-the-ukraine-war-killed-off-the-ground-attack-aircraft.

[14] Alexandra Stickings, 'Space as an Operational Domain: What Next for NATO?', RUSI Occasional Paper, 15 October 2020, https://rusi.org/explore-our-research/publications/rusi-newsbrief/space-operational-domain-what-next-nato.

actors' capabilities in these areas. For the purposes of this primer, electronic warfare in all its forms is an increasingly democratised capability that will continue to shape warfare over the coming two decades. It is also a *developing* capability, making precise effects and reach difficult for adversaries to define, and consequently presenting a challenge to the identification of the elements that are relevant to battlecraft (and therefore the direction of norms) over the timeline of this primer. Extraordinarily, any actor with a computer now poses potential threat. This is certainly a new norm where sudden, unobserved and unexpected effects can be wrought by state, non-state, individual and rogue parties alike. Cyber operations, however, are difficult to undertake. Offensive cyber capabilities require comprehensive workforces incorporating liaison officers, strategists, coordinators, targeteers, lawyers, trainers, technical analysts to process information during and after operations, technical behaviourists to calculate likely responses as well as remote operatives embedded in target parties.

In this vein, NATO's prioritising the primacy of electro-magnetic access is recognition of the domain's new importance.[15] Indeed, this project's initial evidence base returned again and again to the imperative of training for joint operations in electronically contested environments. The issue (and prize) for policymakers is the tricky priority of ensuring resilience and protection of critical national infrastructure against such sinuous threat. Various norms therefore arise. It is, for instance, the increasing ubiquity of data in military and civilian processes that provides unpredictable and novel targets for adversarial activity, exacerbated by a constantly evolving palette of new tools for those actors to outwit countermeasures.

Developments in cyber effects are also transforming this space, making it an arena that is difficult for policymakers to navigate. Cyber's workings are supersensitive, and this further obfuscates any general understanding of the technology's capabilities. It also frustrates any audit of its potential in battlecraft. Given the dual commercial and military ramifications, its deployment must remain strategic, but this also is problematic given the imprecision, long lead times and target specificity of its tasking. Changes in norms therefore remain balanced between high (although imprecise) impact for a successful engagement and the currently low probability of an attack reaping expected results. Cyber operations, after all, involve

---

[15] NATO Bulletin, 'NATO Electronic Warfare Advisory Committee Convenes in Brussels', NATO Publications, 25 November 2019, https://www.nato.int/cps/en/natolive/news_171280.htm.

the crafting of complex payloads without being detected. The norm remains that each attack plan is prone to obsolescence. It is also difficult to synchronise cyber action with an actor's wider plans and that action must currently take place without relevant doctrine to define its deployment. Its impact then depends upon an adversary's readiness, the extent and value of countermeasures, training and response waterfalls in place as well as levels of luck and exogeneity that still contribute to such an attack's success. The challenge for norms is that precedent is not yet a reliable indicator as parties' fielding capabilities have generally been directing attacks against lightly defended targets.

The first year of Russia's invasion of Ukraine demonstrates that broad use of cyber is limited by several challenges and these together constitute the current norm set for this means of attack. First, it has poor signalling effectiveness. Given its transitory nature, cyber is obviously at its most useful when accompanied by secrecy and deception and this, by definition, conflicts with any signalling role. Furthermore, despite its apparent proliferation, cyber requires a lengthy process and one which is very expensive to undertake. It may also attract unwelcome second order effects. These are difficult to forecast and complicate norms should their provenance be discovered. Furthermore, both domestic public opinion and conflict management are muddied by activities in this domain. There is a risk of triggering miscalculation, the more so given that computer networks tend to control an adversary's most destructive weaponry. Finally to this point, the prevailing norm remains that cyber activities risk escalating responses from one theatre to another and likely move actors to the use of military force.

Its verso is that modern conflict is a fight for human minds as well as for their technology.[16] Cyber activities targeting an adversary's public and troop morale therefore fit this mission. Russia's information war is a case in point with its insinuations that President Zelenskyy had fled the country early in its campaign. Indeed, cyber-enabled influence operations to micro-targeting individuals with tailored messages. The norm here is that they should resonate with their targets' prior beliefs and have successfully been undertaken, it is assumed, both in recent American elections and during UK's Brexit process.

---

[16] Jelena Vicic and Rupal Mehta, 'Why Russian Cyber Dogs Have Mostly Failed to Bark', *War on the Rocks*, 14 March 2022, https://warontherocks.com/2022/03/why-cyber-dogs-have-mostly-failed-to-bark.

8
Change Agents in Behavioural Norms

Each flex in norms that is identified in this primer should inform in some way the practice of military leadership and how future commanders undertake their trade. Indeed, several of the behavioural trends identified in this primer posit foundational change to how battle is fought and, taken together, suggest that material adaptions are likely in the period addressed by this book. Norm divergence and development generally mirror the emergence of one or more discontinuities in practice, examples here including the deployment of certain new battlefield technologies, the incremental replacement of human supervision in military tasks, the introduction of cognitive and physical improvements to human conduct as well, perhaps, as human-computer interactions that promise to maximise battlefield performance.

Leadership and People: Future Command and Control Structures

Enduring normative transformation, however, relies upon human leadership if these developments are to be integrated and prove long lasting. Leadership also remains the key to dealing with surprises arising from inevitable second order effects, the more so as big data and artificial intelligence infuse operations and upend how commanders understand

and then leverage risk. This, moreover, is becoming ever more important (and therefore to be reflected in ongoing norms) as response times shorten and those making decisions deal with an exponential rise in the number of available data points and more unpredictability in operations. In doing so, leaders must construct, preach and live with correspondingly foundational changes in the prosecution of warfighting including changes to conflict's organisational, ethical and legal aspects and to do this both from a frontline but also doctrinal perspective.

Leadership and Norm Development

Leadership is often noted as an army's least expensive resource but also its most expensive determinant of outcomes. In considering norms, this adage is worth analysis. The principal change agent here relates to the degree of intervention that humans will retain in war's many loops. Erosion of this link is already evident at a weapon level, but norms must now factor for an increasing reliance upon machine capabilities in order for commanders to undertake their trade. Indeed, the unsurprising norm here is that technology has several points of impact in the practice of command over the timeline of this primer. Commanders must factor for a persistent decline in manning levels, enduringly driven by political and budgetary factors.

The phenomenon is also suggested by that same incremental substitution of human for technology that is discussed throughout this primer. Humans require sleep, complex casualty evacuation and pensions. Their lot is influenced by an uncomfortable but inescapable political dimension, whether that be from an aversion for body bags (indeed, can it be ethical to deploy human soldiery when machines can complete similar tasks relatively as well?) or from the scarce resources required to keep soldiery trained and content. Humans, moreover, are poor compared to machines for dangerous, deep, repetitive or dirty tasks and, as technology develops, it is this unstoppable shift in relative utility that is increasingly norm changing. Unstoppable, that is, until that technology breaks and leaders must lead their assets using more tried and tested skills that have characterised the nature of war since conflicts first began.

Russia's invasion of Ukraine suggests the need to revisit certain assumptions. The long-dated notion that cutting off the head of the Hydra will similarly decapitate a country's armies clearly does not stand up to scrutiny. The effect on Russia's warfighting of Ukraine's precision strikes targeting 14-or-so of its frontline generals during 2022 has been quite muted; the enduring norm is that lower-level commanders always step up and assume new responsibilities and, anyway, an army without a head is still going to fight regardless of the labels that theorists might append to particular actions. Rather, it might be wars of attrition and the resulting commander now denuded of any forces to allocate that might be the genuine decider here. Destroying command nodes and causing paralysis in the adversary's battleplan may appear front footed but evidence suggests that it likely remains the slow, prodding approach which grindingly removes an adversary's means to fight that is the way to victory.

Substituting humans with machinery has other norm ramifications. First, 'old laws' are empirically less precise and more difficult to apply to incoming new technology. Second, it is soft and tangential factors that really drive leadership, service and duty. Leadership is tied into training, what to expect from soldiers, how to motivate and inspire and cajole. It is the art of persuasion, often in the most difficult of circumstances. Like context and ambiguity, this cannot be coded for and, as such, represents a key and immutable norm upon which much of this primer's findings are linked. Leadership is everything. The normative behaviour of organisations exists because the entire chain of that grouping understands and is governed by a set of behaviours and expectations that is the basis of this primer. Norms are about people and leaders lead people. This also acts as a brake on wholesale change (indeed, the incidences of runaway norms are very few) but, in terms of their leadership, groups also rely upon other boundaries, for example a clear strategic framework. Groups, after all, are unsurprisingly weakened by inconsistencies in frameworks, poor readiness or poor strategic design.

The relationship between leadership and norms also works in reverse. Pressures upon norms individually, collectively and cumulatively then act as change agents to leaders and how they practice their craft. It is a virtuous circle. Similarly, norms of warfare are informed by particular styles and types of leading. This has profound implications for an army's ability to adapt and change 'at the speed of relevance'. The implication here is that parties risk falling ever further behind unencumbered potential adversaries whether because of regulation (organisational rules), norms (accepted ways of doing things) or cognition (leaders' individual processes). It is therefore the meld of these three models that will determine how leadership norms will develop going forward. Adjacent developments in social media and connectivity then affect leadership norms in the management and control of public buy-in, public support and the intricacies of maintaining 'hearts and minds'. Similarly disruptive are changes in managing morale and associated issues around permissions and performance.

Other intangibles affect ongoing norms in this space. Commentators, for instance, occasionally note that an 'anti-intellectualism' remains at work in British Army circles. There is no reliable way of triangulating this and its inference that war's norms are more shaped by societal and civilian customs. Certainly, the primer's evidence base suggests that norms are more affected by the contours of political landscape rather than by usual military 'narrowing down' factors that have traditionally influenced norm development across the services. A further leadership norm arises from a perception (accelerated by the world's connectivity) of a pervasive 'failure of experts' in recent event and conflicts.

Experience, Training and Deployment Norms

At an application level, defence reviews have often been more about equipment and less about human resources, training and organising. The alleged discrepancy has not been helped by an increasing vacuum between leaders' lack of in-field experience and battlecraft tenets as reflected in states' doctrine. This mix, moreover, may also be further complicated by ongoing contributions from scholars and input from think tanks. In considering new norms, therefore, consideration is required not to parse battlecraft's increasing complexity with simple unfamiliarity.

For this reason, norms around command and leadership tend generally to focus upon the 'key effort'. This is then supported by other soft initiatives such as the fostering of initiative, the care of soldiers and their wider families, and making relevant training of those troops' first priority. The norms of leadership and effective training are based upon the aphorism of 'growing upwards and teaching downwards', this is undertaken in an effort to encourage innovative leaders to link new explanations to old problems. It is also about embedding the art of restraint, the ability to listen and, in battle, an ability to understand failure, embrace surprise and leverage chaos. The norm here is that it is adaptive, creative and resilient leaders who can minimise internal friction while maximising the enemy's friction. The leader's role is about information sharing and ensuring information flows in times of bottlenecks when information is an overt tool of empowerment. Modern leadership in UK forces is less about overt command and control and, instead, fosters the delegation of authority. The norm here is that it embraces variety and different ideas as is foundationally based on individual commanders' deep understanding of the military profession, the culture of the nation as well as the organisation which that leader serves.

While it is difficult to make more than preliminary observations on lessons to be learned from the Ukrainian conflict, certain systemic weaknesses continue to shape battlefield norms in the spaces considered in this chapter. An example might be Russia's seeming inability to undertake appropriate battle damage assessment. This is interesting as it points to wider weaknesses that themselves constitute a passing norm across parties. Russian military practice appears to presume that if an action has been ordered and carried out, then it has succeeded unless there was direct evidence to the contrary. Here, evidence of success appears to have been based upon friendly parties' confirmation that targets had been destroyed, confirmation from space assets showing damage, as well as evidence that the opposition had themselves reported both strike and damage to their equipment. After all, poor practice in such damage assessment makes parties highly vulnerable to deception. Presumption of success tends to prompt that party to take unjustifiable risk in its force disposition, whether that be lack of support, inappropriate assumptions around the adversary's strength, or misplaced belief in one's own tactics, logistics and disposition. In the case of Russia, command appeared to treat all Battle Tactical Groups (BTGs) as being comparable units of action with new tailoring of tasks to their respective capabilities resulting in tasks being assigned for which units may have been particularly poorly equipped to carry out.[1] Confusion then as to the composition of units combined with changes in pre-assigned tasks would then lead to command decisions being paralysed given poor general situational awareness.

In considering leadership norms, it is useful to keep one eye on Russian control and management. A norm-relevant observation arises, for instance, from their forces' tendency to prosecute orders long after it has become apparent that assumptions in those orders are incorrect.[2] Continuous attempts to assault Bakhmut long after it had ceased to be Russia's main effort demonstrates that, until an order is countermanded, commanders will attempt to execute their last instruction. A further aspect of Russian orders continues to be the near absence of reversionary courses

---

[1]   Mykhaylo Zabrodskyi and others, 'Preliminary Lessons in Conventional Warfighting from Russia's Invasion of Ukraine: February-July 2022', Royal United Services Institute, December 2023, 30.

[2]   Ibid, 47.

of action. If unsuccessful, or if the higher intent is no longer achievable, this leads to requests for clarification being referred upwards.[3] Furthermore, all reported contacts are generally treated as true. Similarly, all fire missions appear to be given equal priority and are prosecuted in the order in which they are received unless an order to the contrary from higher authority. The norm here is that those directing Russian fire missions either do not have access to contextual information or are indifferent to it.

Finally to this point, a combination of too few experienced tactical commanders and a culture that places little importance on contextual judgement leads, suggests RUSI's 2022 report on preliminary lessons from the conventional warfighting undertaken in Ukraine, to Russian forces being systemically vulnerable to deception methods. Indeed, the observation is *generally* relevant to reviewing battlefield norms and arises from the tendency in Russian forces to treat information received as true, to discourage honest reporting of failures and to perform inadequate damage assessment. It also arises from a Russian inclination to design systems around single missions. Even within an electronic warfare or air defence system, each operator controls a specific and different sensor or function. Russian operators are trained to focus upon the specific picture for which they are responsible. The norm here would appear that there is little effective fusion process with little opportunity for incentive to identify inconsistencies within an operationally relevant timeframe.[4]

Fluid Operational Design

A further theme from this project's evidence base is then that the study of warfare is either broken or undertaken with such poor rigour to inform policy that analysis of warfighting is an increasingly lonely trade. The inference here is that fewer parties are interested today in the deep analysis of battlecraft's application and, particularly, its logistics. While not a change in norms, the reasoning is that commentators can only opine on future warfare if they are familiar across current warfare, force capabilities and shortfalls. The issue is that decision processes about how armies will fight are ever less transparent and ever less understandable. This may appear strange in a new era of big data and ruthless collation of digital information. In considering context and norms around the mechanics of

---

[3]   Ibid, 48.

[4]   Ibid, 49.

war, the dilemma is best evidenced in the dynamically changing equation between think tanks and academia versus observers with familiarity and experience who can call on empirics in making informed decisions.

The ongoing Russo-Ukrainian war calls into question deeply held military axioms on force design and campaigning structures.[5] Indeed, Russia's performance in the war's early phases speaks loudly on what can happen when design and structures mismatch politicians' aims and, of course, the empirics of what is actually happening on the ground. Two observations for current norms arise. The first, notes Johnson, is the misplaced belief that future wars will be short, decisive affairs. This has several profound ramifications to norms and the composition, management, and sizing of battlefield assets. If, after all, future conflicts with peers are all protracted and involve significant attrition, can countries with relatively small, all-volunteer armies and no ready and robust personnel replacement systems prevail?

Second, force structures for protracted campaigns look very different from those intended for short, sharp struggle. The longer the campaign, the more embedded resilience, buffer, support and reinforcement must be in commanders' plans. Similarly, longer campaigns require very different pacing in order to preserve forces while still maintaining appropriate tempo. An emerging norm is that the West may be witnessing the end of short wars between states by professional armies. Instead, Russia's playbook in Ukraine suggests a grinding war of attrition taking place primarily on land and over contested territory that both sides covet, strengthening the resolve of combatants.

With regard to campaign structures, war's duration also has consequences for casualties and their replacement, force preservation and unit reconstitution as well as resupply, recruitment and training. Long wars require that populations be engaged, informed and on side. Narratives, fatigue and buy-in require management and investment. Protracted fights also mean high levels of materiel use and waste. In these circumstances, it is no longer all about the decisive winning of any first battle. Instead, it is the maximum number of forces which is available at particular times and at a deployable level of readiness that matches the very worst scenario planning considered realistic by the commander. The norm here is that this elongates

---

[5] David Johnson, 'A Modern-Day Frederick the Great? The End of Short, Sharp Wars', *War on the Rocks*, 5 July 2022, https://warontherocks.com/2022/07/a-modern-day-frederick-the-great-the-end-of-short-sharp-wars.

processes, introduces uncertainty, worries home-grown communities and generally complicates battle management.

It also conflicts with other long-dated norms, including that these commanders are only prepared to fight the war they want and not the war that the enemy has visited upon them. A resurrected norm from the Ukraine conflict then relates to weaknesses around logistics chains, especially in relation to general replenishment of supplies, of human resources and materiel. Logistic pinch points are also exacerbated by civilian vulnerability to cyberattack and disruption. Just as high readiness should be a measure of army force design, resilience and available buffer in its logistics chain are similarly important indicators of preparedness.

Resilience Norms, Redux

The evolving norm here is that resilience and logistics remain key decision variables to adversaries, both in terms of identifying vulnerabilities and discerning strength and in terms of detecting opportunity for new attack surfaces. Here, this project's evidence notes that the UK and its Western allies are at an increasing disadvantage: ethical constraints more generally affect Western responses than more autocratic adversaries. This openness has other consequences. In a similar vein, the training pace and tempo of Western allies are increasingly restricted by health and safety constraints under which several constituent parties must operate, including minutiae such as range safety and ammunition limitations, restrictions placed upon drivers' hours and, generally, the state's willingness to accept training accidents and failure, the cumulative effect of which is to compromise readiness and response.

In reviewing norms around appropriate force design, it is key to remind oneself what makes an army useful. Current norms focus upon decision-making that depends upon levels and responses to threat, deployment of appropriate forces and then being able to sustain that battle at range. A useful army is also about being highly lethal in a manner that also understands the 'how' of fighting which factors for logistics, movement, supply and operating in super-dispersed fashion. In this vein, peripheral assignments for that army such as humanitarian tasking can only dilute these norms given, after all, the limited number of training days available to structures each year. Similarly, those norms must account for the conducting of mobile campaigns over large areas and at long distance and the challenges that this entails around movement and speed as well as what constitute acceptable operational comprises. This also needs to factor for available training days, costs and benefits. Norms therefore need to reflect a new importance of distinguishing between 'complexity' and plain 'unfamiliarity'.

The Ukrainian conflict has unsurprisingly highlighted the importance of sound logistics. The largest challenge during the conflict's earlier stages was to manage equipment losses and the extraordinary expenditure of ammunition. Although Western support had been symbolically invaluable to the Ukrainian army, Russia's invasion was blunted principally through the employment of Ukrainian arms.[6] Newly delivered arms became inoperable due to inadequate maintenance and misuse by inexperienced crews. Piecemeal delivery of a very wide range of weapon systems proved particularly problematic for integrating heavy weapons such as artillery. The breadth of incoming equipment also meant that maintenance and logistical constraints further complicated how such systems were deployed. Also affecting norms were the very high attrition rates experienced by certain weapon categories. Of all uncrewed aerial vehicles (UAVs) used by Ukraine in the first six months of the conflict, some 90 per cent of deployed assets were subsequently destroyed.

As discussed in the previous chapter, the average life expectancy of a quadcopter remained around just three flights. The average life expectancy of a fixed-wing UAV was just six flights.[7] Irrespective of efforts to pre-programme flightpaths, the use of terrain to shield uncrewed systems and to limit substantially data transmission from these assets, it has proved difficult to extend the survivability of UAVs. Moreover, protective measures often reduce these assets' effectiveness in the field. Flying uncrewed vehicles on mute weakens timely target acquisition before the enemy can displace. Furthermore, defensive measures, while improving survivability, also requires precise locations to be selected *in advance* of a mission. UAV missions then fail because there is no target at such specified locations, because missions are disrupted through adversarial use of electronic warfare, the dazzling of its systems or through enemy disruption of navigational systems.

---

[6] Mykhaylo Zabrodskyi and others, 'Preliminary Lessons in Conventional Warfighting from Russia's Invasion of Ukraine: February-July 2022', Royal United Services Institute, December 2023, 36.

[7] Ibid, 37.

Human Rights and Civil Society Norms

Russia's tactics in Ukraine provides much evidence of random attacks on random places, far from the front lines and without military significance. Two considerations arise. First is the deliberate targeting of civilian and non-military assets to puncture morale, sow chaos and complicate governmental effectiveness. It also evidences a new norm in warfighting. Unlike Ukraine's eastern territories where soldiers fight on either side of a discernible front line, elsewhere the country has witnessed tactics that less resemble war and look instead like multiple acts of terrorism.[8] There is no heed paid to the laws of military necessity or proportionality. As Applebaum points out, Russia's lack of traditional war aims combined with the blunt application of force designed to intimidate civilians and influence government policy through coercion appears much like terrorism. It deliberately blurs any distinction between state-administered terrorism, military war crimes and other non-attributable actions designed to terrorise populations, demonstrate contempt for global institutions and, in so doing, show disdain for legal frameworks and accepted norms, a second order effect being further attenuation of international law and practices that are designed to prevent such acts. Indeed, Appelbaum finds it necessary to note that Russian actions in Ukraine are also 'targeting the values that lie behind [such structures], the principles and even the emotions that led people to create them in the first place… All of these moral assumptions have been cast aside by an army determined to create pointless, cruel, individual tragedies one after the next'. While this may appear simply to slide parties back to earlier norms of warfare's brutal nature, undoing at a stroke much of institutions' recent refining of codes of conduct, it actually greatly affects war's recently presumed norms and practices.

Several of the likely scenarios from the Ukrainian war have direct ramifications on global human rights. First, the invasion is causing a very significant humanitarian crisis with ongoing and immediate effect on norms generally. By the end of 2022, nearly six million refugees had fled Ukraine with another eight million having left home to seek shelter elsewhere in the country. This massive movement of people is itself norm-forming and of course affects both how war is undertaken and the effects it wreaks.

---

[8] Anne Applebaum, 'Russia's War Against Ukraine Has Turned Into Terrorism', *The Atlantic*, 13 July 2022, https://www.theatlantic.com/ideas/archive/2022/07/russia-war-crimes-terrorism-definition/670500.

Similarly, it reinforces the long-dated norm that it is the most vulnerable who suffer from war, especially those at the base of the Maslow hierarchy of needs (those requiring food, warmth and shelter). That energy policy is rotating quickly towards secure access and source diversification also impacts war aims and, in time, how warfare and political aims are shaped. Indeed, norms again need to factor for food security and parties' race for critical materials, for key equipment and the commodities with which to manufacture such materiel. Indeed, in terms of military procurement, the invasion has engendered a new era of supply chain management and a shift from just-in-time to just-in-case manufacture.

This in itself must be an emerging norm of warfare. Moreover, an adjunct norm arising from the Ukrainian conflict's first year is that global standards are now more likely to splinter on partisan, bipolar lines and to do this with considerable societal ramifications. In this vein, the war has generally accelerated a previous trend for certain countries to cordon off a wide range of online content services, limiting what residents can see and do. This is likely to be a continuing behaviour, considerably affecting human rights in those territories. Lastly, hardware standards, internet protocols and ongoing engagement by corporates in those countries involved in conflict is more generally likely to change as a result of Russia's action towards its neighbour. In the round, therefore, human rights and the conditions in which humans live in conflict zones will continue, as always, to be adversely affected by resulting instability, whether from governmental constraints, from food and fuel poverty, from volatility in financial systems, or from the more parlous state of state finances occasioned by the rise in defence spending or by increased cyber activity by adversarial actors.

It has been recognised that the conduct of war has broadened from what had been formerly considered as military business. Much of the foundational evidence of this primer speaks to activity below the acknowledged threshold of war, whether in the cyber, espionage, economic or information domains. In line with this, those undertaking warfare must consider the implications arising from broader communities' involvement in each conflict, quite apart from those who take up arms. Commentators still note the difficulties of navigating wars when combatants do not wear uniforms and clear insignia, despite this having been a longstanding feature of warfare. Insurgencies meet this description and have been part of military canon for generations. However, a developing norm is that planners must factor for a much more vocal, influential civilian involvement around military activities. Indeed, the role of civil society is inextricably linked to the conduct of war from those supporting efforts through factory work and agriculture, to lawyers, activists and, more challenging, those civilians who find themselves as targets of war despite ostensible protection enshrined in international humanitarian law. While modern warfare may be closing gaps between industrial support and combatants, current conflicts only reinforce this norm.

A further development here is the growing importance of unarmed civilian contractors in military forces, whether they be involved in the repair of analogue equipment from rifles to combustion engines or, more likely in today's force structures, the configuration, integration and then maintenance of very sophisticated electronics. Nevertheless, keeping vehicles and weapons in play is crucial to all theories of victory. The war in Ukraine has shown that wear and tear of hardware has far outweighed projections. After just ten months of fighting, assessments in Ukraine put the combined loss of platforms at over 11,000, a figure comprising tanks, fighter jets, helicopters, infantry fighting vehicles and artillery pieces.[9] These figures dwarf the entire stocks of NATO countries. At the start of 2023, the UK, for instance, had 227 Challenger tanks. France recorded just 628 VBCI infantry fighting vehicles. Germany had 38 MARS multiple launch rocket systems. Whilst some vehicles may be damaged beyond repair, others produced in the Soviet era may be relatively simple to fix. The developing norm, however, is that modern military equipment is vastly more complicated, composed more of computer code than nuts and bolts and beyond the skillset of military mechanics. Indeed, in many documented cases, soldiers are not allowed to repair their own equipment.[10] They are replaced instead by cohorts of civilian contractors deployed forward in order to keep complex equipment running. During the war in Afghanistan, for instance, it was civilian mechanics who were based in Camp Bastion who repaired equipment in forward positions rather than that equipment returning to the UK.[11] In this vein, platforms such as uncrewed aerial systems as well as modern crewed platforms like F-35 fighter jet or the AH-64E attack helicopter are routinely accompanied by civilian field service representatives who are technically adept and able to reach back to the manufacturing parent company in order to coordinate software fixes

---

[9]   Richard Thomas, 'Russo-Ukraine war equipment loss ten times that of Moscow's Chechen conflicts', *Army Technology*, 23 December 2022, https://www.army-technology.com/features/russo-ukraine-war-equipment-loss-ten-times-that-of-moscows-chechen-conflicts.

[10]   David Dayen, 'When Big Business Won't Let the Troops Repair Their Equipment', *Prospect*, 19 September 2019.

[11]   Ministry of Defence, 'Civilian mechanics keep wheels turning at Bastion', *Ministry of Defence*, 4 November 2011.

and manage troubleshooting. The developing norm is therefore that such civilian employees close to the front line have 'a direct and mission-critical support function', a new but not particularly well understood paradigm.[12]

Civilian software engineers have similarly been a critical part of the war effort in Ukraine. An example is useful. Ukraine has leveraged its position as a leader in software development to develop GIS Arta, an artillery targeting application (Geographic Information System for Artillery).[13] Ukrainian soldiers can pass information about suspected targets via the application to a command and control node which then requests fire from a suitable and available element. This may be a mortar crew, artillery or a precision strike. Commanders on gun lines can monitor the app and engage targets that are highlighted. The app enables a rapid targeting cycle with great accuracy.[14] This civilian-military interaction is a development from traditional military-industrial relationships that had been the preserve of defence 'primes' such as Boeing and Lockheed Martin. A similar step change from previous practice can be seen in the civilian contractors employed by the United States Department of Defense to fly its MQ-9 Reapers in an intelligence collection capacity (rather than strike) since 2015.[15] In Ukraine, moreover, a network of largely unsupervised civilians use their own vehicles to transport people and vehicles around the battlefield. Military fuel bowsers which provide obvious targets for an adversary have been swapped for civilian equivalents. Civilian metal smiths have been responsible for constructing war materiel from anti-tank obstacles to basic portable stoves. Backpack manufacturers have made body armour plates; volunteers have filled sandbags to protect urban infrastructure while others have used churches to make camouflage nets. It has largely been left to civilians to document war crimes or work with local businesses to support and reinforce supply chains.[16]

---

[12] Andreas Wenger and Simon Mason, 'The Growing Importance of Civilians in Armed Conflict', *CSS Analyses in Security Policy*, Vol. 3, No. 45, 2008, p. 2.
[13] Mark Bruno, '"Uber for Artillery"–What is Ukraine's GIS Arta System?', *The Moloch*, 24 August 2022, https://themoloch.com/conflict/uber-for-artillery-what-is-ukraines-gis-arta-system.
[14] Halyna Kubiv, 'Gis Arta – So kann eine Karten-App den Krieg in der Ukraine entscheiden', *Macwelt*, 9 June 2022, https://www.macwelt.de/article/989799/gis-arta-so-kann-eine-karten-app-den-krieg-in-der-ukraine-entscheiden.html.
[15] WJ Hennigan, 'Air Force hires civilian drone pilots for combat patrols; critics question legality', *Los Angeles Times*, 27 November 2015.
[16] Oleksandr Sushko, 'Defending civil society in Ukraine', Open Society Foundations, 8 March 2022, https://www.opensocietyfoundations.org/voices/defending-civil-society-in-ukraine.

Crowdfunding of equipment has been a further albeit related development. While public fundraising has been a characteristic of war for centuries (war bonds and the like), the war in Ukraine has seen a step change in the practice, accelerated in the main by the internet. From sponsoring messages on shells intended for Russian lines or selling keyrings made of downed Russian fighter aircraft, civilian ingenuity has spiralled. Much fundraising has been focused on consumable goods like clothing and sleeping bags, but $20 million was notably raised to purchase TB2 remotely piloted air systems, night vision equipment and sniper scopes.[17] The practice can complicate procurement, suggesting perhaps that official channels cannot meet the needs of their own soldiers. A verso, however, is that is also provides a useful hook to generate interest within civilian communities and demonstrate to soldiers that they have support beyond their central military establishment. Ukraine, moreover, has been practising for its conflict with Russia at least since 2014, and not February 2022, dealing with disruption both in the physical and cyber realm for nearly a decade. It cannot be assumed that civil society is the West is as well attuned to potential conflict. Such planning conditioning takes generations to develop.

---

[17] Andrius Sytas, 'Lithuania to transfer a crowdfunded Bayraktar drone to Ukraine on Wednesday', *Reuters*, 6 July 2022.

Conclusions

War and warfare have long taken up authors' energy, particularly around how the two conditions will change and evolve in the future. What is also observable is that none of these pasts, presents or futures have come to pass on the battlefield. Instead, combat experiences retold from Iraq, Afghanistan, Georgia, Syria, Ukraine and Israel (amongst others) illustrate how militaries face different challenges and realities from those posited by these commentators. But this was as true for Romans after reading Heroditus as it is for Ukrainians reading Singer.

In this vein, the research interviews that constituted this book's primary evidence drew on a broad cohort across the defence community between 2019 and 2021. The authors' intention was to use this extraordinary resource to provide foundational context to this primer. Without exception, however, those interviewees, a curated and undoubtedly expert body on war and warfare, pointed generally towards a future of war as an extension of present means, a linear progression of activities based upon strategies and tactics that have been used by adversaries in recent campaigns. In this sense, the primer's initial context proved unexpectedly homogenous. And wrong. Asked to look forward and propose likely norms and forms of warfare 15 years hence, participants (perhaps understandably) suggested pathways anchored by hybrid, sub-threshold and other politically framed models of warfare, all enabled by technological progress, by autonomy and means around plausible deniability and all buttressed by envisaged developments in the domains of cyber and space warfare. This may, of course, yet prove to be prescient but does not reflect Russia's tactics in war in Ukraine. To be an expert, after all, one's forecasts must be better than chance.

With the easy benefit of hindsight, Ukraine's experiences since 2022 demonstrate that this set of expected forms and norms has just not played out. Rather, predictions have been far too closely informed by Western experiences in campaigns across the Middle East, North Africa and Eastern Europe between 2001 and 2021, the seeming eclipse of conventional warfare's golden era and, frankly, the end of massed and brutal battles between military forces. Indeed, the end of the Cold War was widely heralded as a whole new epoch of how new war would be undertaken. As with this book's thoroughly reductive treatment, this was the broad narrative into which RUSI launched its project on war's new norms and also the footing from which this book now emerges. In considering the balance, moreover, between conventional and hybrid means, forecasters of war's processes had instead drifted towards quite binary either-or scenarios rather than staying anchored by an equilibrium based upon a *portfolio* of means. This is an important observation as, eight chapters later, this primer situates itself squarely in this *portfolio* camp; clever new strategies are rarely new and their effects rarely revolutionary, and they will anyway still retain a conventional component whether that cleverness comprises the several hybrid means discussed throughout the book or not. The same holds true, less granularly, around overt and state-sponsored initiatives such as those of China, its Belt and Road Initiative and its mix of legal, psychological and media activities that together make up its 3 *Warfares* doctrine.

In considering, therefore, where we are today, it becomes disproportionately useful as 2023 draws to a close to be able to review these discrepancies. A case in point is obviously provided by Russia's ongoing actions in Ukraine. A key exercise, after all, for this book has been to consider empirical developments on that battlefield versus the portfolio of outcomes and practices that were being conjectured *before* Russia embarked upon the invasion of its neighbour. Indeed, understanding the degree to which anticipated forms of war have diverged from those being experienced on the ground was an early aim of the authors' analysis, a consideration of the extent to which methods and forms of warfare have actually morphed.

In this first instance, our commentators judged that anticipated forms were going to blend not just between domains of warfare (air, land, sea, space and cyber) but also across environments (military, political, diplomatic, economic and informational). To this end, readers should not have been surprised with the playbooks of the time, all considerably based upon Western leaders' adoption of a quite single set of scenarios, primarily driven by a leading role afforded to technology and the rise and rise of its role in warfare.

Those playbooks, moreover, had predictable ramifications. Militaries (and uniformed humans), it seemed to all, would play a less important role. Instead, leaders in Europe were placing greater interest in recruiting 'pentaphibians', coders and spy masters rather than weapon operators and soldiers, sailors and aviators. And military investment decisions were already reflecting this trend. Budget allocations were increasingly being pushed towards 'new' technologies and away from conventional platforms (tanks, fighting vehicles, ammunition stocks and the like). By 2021, it looked very much that political leaders had initiated a decade-long reset of their militaries in order to equip for exactly these new forms with which to undertake conflict.

Russia's invasion of Ukraine in February 2022 has radically altered narratives about war and warfare in the West. Russia's use of a decidedly mixed-doctrine playbook involved a quite different form of war than had been expected. To a considerable extent, it was a plan inspired by US military successes in 1996 and 2003 when 'shock and awe', lightening raids and regime change drove the wider agenda. It was also, fortunately for Russia, backed up by its more traditional form of war: attritional, positional, and soon to be focused on material destruction. Whilst the original plan predicted quick success and regime change in Kyiv involving special forces, the Kremlin's general staff always planned on using divisions of their conventional military to 'fix' the most capable Ukrainian fighting forces in the east of the country. This was clearly a departure from the expectations of Western intelligence analysts, military leaders and politicians and was essentially a replay of Russia's 2014 annexation of Crimea and Ukraine's eastern areas, albeit accompanied by intense political and economic posturing. A heavy cyber presence had also been expected, involving the shut-down of the Ukrainian power and water grids as well as widespread denial-of-service attacks designed to paralyse the machinery of government, trade and communication. On this point, however, readers should note that neither cyber sabotage nor cyber misinformation really amount to much-touted cyber war.

What subsequently unfolded was therefore roundly unexpected in Western capitals: the lack of an effective cyberattack by Russia; a failed regime change and the rapid withdrawal of Russian forces from Kyiv; heavy bombardment of Ukrainian cities without thought for life or physical damage; the sinking of ships offshore; the failures (on both sides) of logistical management; the eye-watering use of ammunition stockpiles; and, most of all, the shift to conventional forms as the primary means of fighting between the belligerents. This form of warfare would have been more familiar to those in 1915 or 1940 than to people reading policy and political speeches over the last two decades. First, technology was not playing the decisive role as widely imagined and ever more heralded in recent techno-fiction. Second, paralysis from cyberattacks did not occur, although an underlying cyber campaign was undertaken on both sides. Finally to this point, 'little green men' were absent and not at all fundamental to Russian operations.

Western states meanwhile sought to enact considerable economic sanctions against Russia while diplomatic and political measures simultaneously attempted to isolate the aggressor from the international community. And while many Western states have provided weapons, ammunition and training to Ukrainian forces, it has to be noted that overt support for Ukraine in global terms has been muted. Only some 15 per cent of the world's governments agreed to impose sanctions on Russia (this statistic is by population, although in terms of gross domestic product this was far more significant). Nevertheless, tanks, artillery, long-range rockets and anti-tank weapons all at once became the new geo-political currency, with Western democracies falteringly and then quite rapidly offering to bolster Ukraine's defence, much of this equipment and materiel coming from underfunded war stocks and quickly diminishing domestic stockpiles. It is against this backdrop that this primer has sought to identify lessons arising from current conflicts both in the political realm but also in the military domain.

A central enquiry for the primer is the extent to which patterns which were readily identifiable prior to 2022 constitute a long-term trend. In this vein, the issue becomes the extent to which Russia's tactics in invading Ukraine then represent an aberration in warfare's evolution. But this still returns us to the classic 'a war' / 'The War' juxtaposition which faces force designers today just as it has done so many times in the past. In attempting to design a military to meet the challenges of the future, planners must reckon on multi-decade timelines before their programmes bear fruit. In doing so, those same planners must decide whether to base their assumptions on long-term patterns and trends for which there is perhaps cumulative evidence, or instead pivot towards lessons derived from very recent experience. Which is the better indicator for forecasting? This is no simple task as major wars, those that test militaries and states most dangerously, are rare. Moreover, the identification and then learning of lessons is rarely clear and their fallout is expensive to remediate, forcing (as it does) militaries to design themselves around new procurement, new training and logistics, and frictionful infrastructure, and to do so in a manner that likely differs significantly from current practices.

It is not by accident that political decision-makers across history empirically seek areas of common ground between differing force designs, solutions that perhaps have utility in peacetime affairs of constabulary, peacekeeping and counter-insurgency operations but with an ability to swing into full-scale combined arms operation should the existential fight unexpectedly arrive. The approach has merits but it is also one that is dependent on policymakers understanding the exact moment at which they need to recapitalise their forces for high-end conflict, act on that and have plans in place to make those plans real. They cannot be late, regardless of political, budgetary or geo-political factors all throwing grit into that decision. Critically, this includes both their latent industrial bases as well as the societal resilience to withstand difficult initial losses while gearing for the main fight. The process is also fraught with exogeneity: budget constraints, special interests, political leadership, the voice of the enemy. And luck. It is the broad mix of these factors that becomes the lesson and conclusion for this primer.

Historically, timelines in the West for these shifts have been around a decade, the 'ten-year rule' featured in several discourses in the twentieth century as a mean period in which such decisions could be made to reconfigure the economy, industry and society. That schedule was loosely (and empirically) based on some understanding of the challenges involved, the availability of raw materials, the pieces required to reshape the country's manufacturing base, and then the training margin needed to test and deploy these new capabilities into the field. Today, planners must also factor for new frictions, the additional complexities of novel manufacturing and an industrial base no longer able to deliver the required scaling to service the plans envisaged by those decision-makers. A knock-on effect, moreover, is that the construction of military units in the twenty-first century takes a good deal longer than in previous eras. Planners' long-held decision window, the 'decade-long period of opportunity', looks increasingly absurd.

What also emerges from this book's analysis is that, while lessons certainly require policymakers' action, not all of them cost blood or treasure. Decisions are not just about procurement, about buying latest means and pivoting force structures. The accusation is that decision-makers often prefer reflexive, spontaneous pronouncements instead of the considered framing and prioritisation that these often momentous processes require. Addressing this shortcoming might waste less time, save money (and, more importantly, lives) and allow lead-in time to better operational outcomes. These processes, moreover, are long dated for a reason. They require wisdom and courage in decision-making and then steadfastness in implementation. Advance foresight and planning, after all, are the cheap part of this whole equation and only a few of these lessons present as discrete opportunities ('buy this', 'cut that', 'train for everything'). The complicating factor is that most instead cut across this book's several chapters and, as a result, are more difficult to articulate, determine and then navigate.

Various repeating tropes emerge. First, absolute forecasting is difficult. It is also inappropriate. Second, today's febrile pace of change has not really altered that norms empirically evolve slower than might be expected. Warnings against presentism, that tendency to interpret past events in terms of modern values and concepts, intentionally pepper this book. Moreover, while norms may cross over several themes, context ensures that they are certainly not applicable to all situations or to all conflicts. Finally to this point, a further (albeit counter-intuitive) dilemma for readers remains the requirement to test and test again the relevance of *wholly conventional* war in the period under discussion. Can conventional means really win wars? After all, a quite plausible argument can be made that the West has been winning battles but losing wars since 1945: France in Algeria and Indochina, the UK in Palestine and Cyprus, the USSR in Afghanistan, Israel in Lebanon, and, to finish the point, the US in Vietnam, Iraq and also Afghanistan.[1] Selection and confirmation biases are at work when tanks in Ukraine are immediately conflated with World War II, the paradigmatic conventional war, and around Bakhmut where observers are immediately reminded of Verdun and Stalingrad and the notion of 'Win the Big Battle and Win the War'. Indeed, it is fair to note that certain *unconventional* elements of Ukraine generally remain underplayed: Russia's successful use of Spetsnaz, mercenaries and widespread disinformation in its 2014 annexation of Crimea; Ukraine's defeat of Russia's 2022 blitzkrieg using guerilla units firing Javelin anti-tank weapons, images of which were then immediately available across the internet; Russia's subsequent targeting of civilians, flattening cities, its expansion of the Wagner Group's activities; and, finally, the seeming failure of Ukraine's outwardly conventional counter-offensive in the summer of 2023.

This all gives rise to a further signpost. The degree to which forecasting has been consistently wrong (and at almost every level of agency) highlights

---

[1]   Sean McFate, *The New Rules of War: How America Can Win*, HarperColins, 2019. See also Sean McFate, *Goliath: Why the West Doesn't Win Wars, and What We Need to Do About It*, Penguin, 2019.

the pressing need in today's breakneck environment to reexamine the assumptions and heuristics that underpin planners' scenarios. Ukraine alone demonstrates that it has rarely been more necessary to revisit the outcomes that fall from blueprints and preparations. The planner's raison d'être, after all, is not to be too wrong. And the appendix which follows this conclusion (and which sets out in note form the project's first evidence base) is a useful starting point for exactly this reason. The primer, moreover, demonstrates the danger of overlaying one's own models and assumptions onto the actions of adversaries, especially when considering either the costs or timeline of wars. Indeed, a recurring trope from this analysis is that issues around budgets and the prioritising of resources have gained much too much importance in discussions on warfare's forms and norms. What price, after all, is to be paid to counter existential threats? As war becomes seemingly cheaper, conflict might suddenly appear all the more plausible as a credible arm of policy. All of this has ramifications for the planner, even suggesting a lower bar to politicians intent upon achieving their aims, a likely driver in Russia's choice of means in Ukraine. Nevertheless, while current warfare may appear newly complex and fast, the heady mix of collective neophilia (parties' immutable bias towards novel means) together with planners' disposition to consider their moment as a period of unprecedented turbulence (while simultaneously attributing an exaggerated tranquillity to the past) actually points to *slower* evolution in norms and practices since the advent of the information age in the late 1970s. The evidence is that norm change involves a slow, old process.

In all of this our language is also painfully ill-suited to the matter at hand, in particular for discussions around national security. Definitions grate and confuse, alienating soldiers and politicians alike. Audiences are immediately disadvantaged by impenetrable concepts and acronyms. A case in point is our imprecise and increasingly meaningless use of the term 'effects'. Language, moreover, tends to obscure the horrible nature of war, seeding a further concern for the authors whereby the subjects under discussion become so wide and imprecise that they should really elicit the response of 'what were you thinking?' Indeed, the evidence throughout this analysis highlights the inconvenient norm that while we may know more about everything, we actually have less ability either to control or understand what is being presented.

The issue at hand is therefore the credence and authority to give the many change agents that this analysis throws up. After all, the established tradition in how to think about the profession of arms, that the nature of war is constant but that its character is constantly in flux, has been a long-held and expedient construct for those examining the subject. In this vein, the primer's evidence set (its first set of interviews that were started back in 2019) was clearly shaped by what was already a 30-year-or-so narrative of technological progress being the principal driver of performance in warfare. This, after all, is a dominant thread across academic literature and is widely accepted by senior military officers and political leaders alike. But this creates a conundrum. For all this debate over change, over evolution and revolutions in warfare, most of today's battlefields (and, at the date of publication in 2023, there are not less than 28 active conflicts currently being waged around the globe) would remain absolutely familiar to former commanders. The contemporary urban destruction in Ukraine by Russian forces in 2022, or the damage wrought in Syria, Yemen, Sudan, Israel or Iraq would immediately be recognisable to military and political leaders of the past hundred or so years. While Napoleon might have judged Russia's invasion of Ukraine as a textbook example of how *not* to carry out a military campaign, the geography of that war and its destruction of property and targeting of urban infrastructure would be something with which he and his cohort would be absolutely familiar. As an aside, he would have been similarly captivated by the extended range of artillery as well as the opportunity set of drones as new weapons of war.

And it is just as likely that he would view range, precision and technology as something that had merely evolved but was not fundamentally different to how his staff might have undertaken military operations. The enduring norm here is that war remains thoroughly recognisable from that high-level perspective. This translates across battlecraft. Just as the battlefield would retain a certain familiarity to military officers across the millennia, so to would the strategic discussion being held in capitals be familiar to great statespeople. It seems likely that Bismarck, Stalin, Churchill, Roosevelt or indeed Machiavelli would be just as acquainted with discussions on resource allocation, grand strategy, prioritisation, infrastructure protection, war stocks, funding and the dangers of escalation. Deliberations around norm change should reflect this business as usual. Indeed (and perhaps helped in small part by this book's analysis), readers should feel better equipped to reject the neophilia and presentism that exists in much of the narratives today. It *is* possible to judge the degree to which warfare has morphed without being blindsided by individual revolutions that are taking place each day, each week in particular technologies, in new industrial processes and the like.

What then becomes clearer from the evidence to this primer is that much of the change to parties' execution of war and warfare has instead been based on the upending of Western *presumptions* by adversaries and then the divergent behaviours arising among these opposing parties. This is an important nuance from the research conducted over a period of three years for this project. Originally, the research question which the authors considered was 'what will war and warfare look like in 2040?'. Yet the idea that one can construct a future simply by looking at contemporary conflict is as flawed as it has ever been. Indeed, in framing the primer's research and deciding upon the book's chapter headings, the authors unexpectedly found themselves having to confront a set of widely held assumptions and societal biases. Take, for example, the notion that rules properly exist for the international system and for the conduct of warfare. Under examination, this is a bold (and likely erroneous) presumption and at the very least needs a caveat or two. Notions of 'world order', the stuff of conventions and protocols and concords which have been signed from time to time in at the United Nations and elsewhere, are undoubtedly relevant to passing norms but are wobbly at best as the world's scaffolding of that same order.

The uncomfortable question for this primer thus remains the degree to which these treaties and arrangements are just a Western construct, a set of aspirations driven by those with a vested interest in democratic values but with little means of enforcement in the face of quite regular breach. Empirically, moreover, that 'order' is not one that is adhered to by much of the global population. In particular, it is within the paradigm of a rules-based *battlespace* (proportionality, distinction, military necessity and the 'laws' of armed combat discussed in earlier chapters) where infractions have long been particular egregious, leading dispassionate observers to question the validity of the whole construct on the basis of belligerents' behaviour over the last 30 years.[2]

A further observation from the analysis has been parties' inability to learn the required hard lessons from past and present conflicts and, by extension, to understand correctly how norms have moved. There may be reasons for governments and the like either to skew or ignore such deductions. This too is neither new nor revelatory. Indeed, in *The Future of War: A History*, Freeman concludes that there is a peculiar difficulty

---

[2] Peter Roberts and Sidharth Kaushal, 'The Rules of Competition', RUSI, 2020, https://www.rusi.org/explore-our-research/publications/occasional-papers/competitive-advantage-and-rules-persistent-competitions.

in learning those lessons.[3] This, in part, arises from militaries' difficulty with the tricky delineation between 'a war' and 'The War', the difference between a war that is unlikely to be replicated in the future (the analysis of which might be interesting but extraneous) and one which throws up lessons for the future that are sufficiently relevant to merit changes in norms and practices. This clearly chimes with this primer's analysis.

Lessons, after all, must have *longevity* if they are to warrant the expense and pain always involved in adapting and upgrading capabilities. But this is complicated by the divergent nature of the lessons arising from each conflict. Wars throw up the widest variety of recommendations, often contradictory, always dependent upon passing context and usually frustratingly open to interpretation. For example, many Western commentators considered the 2020 use of drones in Nagorno-Karabakh to be 'game-changing'. Yet analysis elsewhere drew entirely different lessons about such capabilities, that their use was instead often restricted by weather and by poor connectivity, providing quite counter-intuitive opportunities to exploit a force's reliance on such assets when, for instance, operations degraded their use, a factor that led to the taking of the high ground in Shusha and the eventual end of that campaign. Drones, moreover, appear to have been quite absent in the parties' subsequent fighting in late 2023. Drone deployment in Ukraine has similarly witnessed extraordinary innovation but deriving sure lessons from their use is very different from concluding that their deployment and countermeasures are required ongoing in some guise. A question for this primer is whether these differences in interpretations are simply a feature of conflict's changing character or whether they point to a new ideation of warfare over the coming decades. Like it or not, we are involuntarily returned to that age-old conundrum faced by planners: Are changes in and between belligerents fleeting or might they permanently alter future practices? Only having reconciled these ambiguities with developments in warfare's forms and norms can those changes in behaviour be understood.

These findings are not new. Absolute forecasting has always been unworkable. Whether because of poor analysis, knee-jerk responses to immediate events, or the selective nature and small universe of case studies, the demand to derive new understanding from each new conflict has unsurprisingly led to poor decisions being made around future force design and military capabilities, all with very long-term consequences. Two

---

[3] Sir George Friedman, 'The Future of War; A History', October 2017, see https://www.youtube.com/watch?v=6xXqrgDP8CA.

observations arise. Again, a contributing complication here is that language being used today, whether for forecasting or for creating policy, remains clearly insufficient to deal with the subject at hand, with the complexity of war and warfare. Here, analysis can be neither reductive nor maximalist. Indeed, we seem to have shied away from the use of complex language, wanting instead to define issues ever more narrowly and so achieve simplicity even as outputs become more and more impenetrable to the lay person. Second, speeches by military political leaders often divine exceptional revolutions in wars and warfare where none exist. Technologies, even new domains, evolve far more slowly than we might think. Take, for example, the idea that cyber warfare would fundamentally alter the way that combat was undertaken, itself a popular narrative between 2010 and 2020, and yet cyber has proved itself to be actually one of the slowest forms of warfare when considered against other options available to decision makers. As the primer's earlier chapters argue, it is not without reason that this new domain of warfare has become neither the dominant nor the decisive factor in armed conflict. The same is likely true around discussions over the importance of space, data, information operations or ubiquitous surveillance. Soldiers, moreover, need to fight when all of these novel (and usually hyper-connected) technologies fail and the evolving norm here should therefore be that expectations will likely disappoint in the cases of biotechnology, nanotechnology, quantum computing and non-kinetic weapons.

Why should this be? A norm might be the generally unfounded optimism within the national security community about what war's new means can properly deliver and the degree and shape of additional capability that can be delivered by these assets. Capabilities, after all, are by definition fleeting and are always dependent upon passing context. Here, the whole matter of 'balance' in force design is ignored at the planner's peril. It is not simply a question of weapons working or failing but whether they are available at all, at what relative cost, and requiring what compromises elsewhere in a party's battlecraft. Today, the expectation is that more can be achieved for much less investment. Unfounded optimism and that same 'revolution in expectation' permeate much of this narrative, neither of these characteristics being remotely new behaviours in the preparation and conduct of conflict. Indeed, while the assumption may be that national security can be achieved ever cheaper and with more utility, the reality remains that more and more data points about every passing variable does little to solve the reality of the problems thrown up in the processes of war. Variables will continue to confound and upend the jobs of those in command, those relying on this data for their conclusions and all staffers tasked with receiving, processing and then interpreting these feeds (the more so given data's worth being limited to particular and fixed moments in time).

Nevertheless, the notion and practices of the 'connected battlefield' are now a staple for parties undertaking conflict notwithstanding that data remains a fractured, exasperatingly fungible commodity. And the constraint remains that commanders are poorly qualified to analyse and opine on its conclusions. The enduring norm is that battlefield data is often wrong, usually partial, certainly duplicatory and soon obsolete. It is, after all, at least as wrong (and, given adversarial meddling, likely more so) as data in all other walks of life. What perhaps has changed, however, is the reality of more universal access to it and the democratised and foundational technologies built upon it. This too has unexpectedly destabilising consequences, whether from the emergence of newly tech-enabled capabilities in insurgent or nonstate groups, or through the destabilising proliferation of sophisticated equipment delivered into the hands of proxies by parties intent on disruption. There is no doubt that the trend is here to stay and requires appropriate flexing in parties' behaviours, albeit anchored to the specific context of each new case.

So, what do we derive from this work? The first and most important conclusion of the book is really that the conduct of war and the execution of warfare have in fact changed quite little regardless of one's timeline. Depending on which side of the fence you sit, this might appear axiomatic or heresy. On one hand (and given recent changes in platforms, in lethality and range and speed which modern weapons can now achieve), commentators erroneously conclude that such rates of change are monumental. And yet on the other hand, artillery remains artillery, and understanding where the enemy is and what it will do remains the most pressing command conundrum. In its broadest sense, war is immutable, and it is where and when and how that commander decides to engage with an adversary that remains the main factor in determining battle outcomes. After all, resources will always remain scarce, and it will always be that militaries going to war will do so without the assets and capabilities to hand that they want to win the fight.

In considering norms, these various axioms are worth reiterating exactly because politicians and military leaders alike appear often to skate round and even reject them. Indeed, the idea that wars can be short and concluded to order should by now have been roundly overturned by the experiences of recent conflicts but column inches and television time will continue to reinforce a deep-seated belief across Western establishments that short sharp military engagements can deliver defined political outcomes. And this will continue to happen notwithstanding examples from the former Republic of Yugoslavia, Kosovo, Sierra Leone, Somalia, East Timor, Sri Lanka, Kashmir, Mali and Libya. War very rarely plays out as predicted by either military academics or political leaders. What instead has been proven is that it is the *will* to fight within a population and its leadership that has far more impact on both a conflict's unfolding and its outcome than the willingness of third parties to intervene whatever the best intentions of those parties. Short war is illusory, an axiom that would be familiar to those in the profession of arms across history but one that seems to have been forgotten in the post-Cold War era.

This prompts a further observation. Changes to the structure of today's forces (and their relative costs) continue to upend how parties will engage in the future fight, not least the perception that there are ever fewer platforms around in parties' arsenals to undertake the battle. But this was as much true for the British Expeditionary Force at the start of World War II as it is for any American forces attempting to deal with China's ambitions in the Indo-Pacific. And, as with virtually all previous wars, the lesson here is that the longer-term success of military action is likely dependent on the war stocks, upon parties' economic base, their logistics chains and (again) the willingness to fight of the parties involved. In 1939, after all, the British Expeditionary Force was more a tripwire and delaying mechanism designed to give British coalition parties an opportunity to galvanise themselves and space to mobilise against the threat from the German alliance. This informs practices today. The time required for that mobilisation was still very considerable despite what had been ten years of political understanding and policy persuading the industries of the Commonwealth to begin preparations in earnest.

Today, those same considerations can be transposed onto Western norms in their measures, for instance, around how to deal with China. While insufficient forces may exist on hand to delay any significant People's Liberation Army advance, it remains that the West's considerable economic and industrial bases can provide an appropriate fighting force at some future moment to thwart Chinese ambition. The difference today, however, is that any such grouping of Western allies starts from a position of being already hollowed out, being brittle and, crucial also for norms, geographically distant from where they need to be. The effects of this upon behaviour is one reason why this primer returns again and again to how militaries will fight, how new means must be integrated into legacy force design and how changes to battlespace must be considered if norms are to be understood. Being late today for the battlefield risks missing the battle completely, the notion of the fait accompli and the subsequent difficulty for parties to unwind an adversary's successful (and usually drastic) action if deterrence has failed and the affronted party is tardy to the initial action. Which, with current states of readiness and materiel, is very likely the case for most Western governments, providing adversaries with time and scope to subvert common fronts and undermine the playbooks that are discussed throughout this book.

The evidence also suggests that the West's current force structures are not designed for high intensity combat and are instead centred on the requirement for a series of constabulary operations. This state of affairs is based around the optimistic belief that their deployment will instead be more akin to a reinforcing role to local partners, and all to be undertaken within this assumed rules-based order that, as above, lives to a large degree in the messaging and language of Western political classes. Nor is the industrial base of the West in any way ready for the scale of recalibration that is required to meet the mobilisation demands of a significant military campaign where one great power is pitted against a peer. That is not to say that such a situation cannot be overcome but, more prosaically in the assessment of norms, it is to highlight scenarios where a considerable fight, potentially on a global scale, has become more likely over the next decade. That lack of preparedness is both a weakness but also an uncomfortable inconsistency when measured against the ambitions and narratives that are articulated by Western politicians.

It is also clear from this inability to predict what the next war will look like that a continuum still exists between, at one end, hybrid menacing to, at the other, high intensity combat. This is a long-dated and dynamic phenomenon and the lesson here is that the procurement of a *balanced* force structure has never been more relevant. The point is worth repeating and, to rather complicatedly paraphrase Howard, given that armies will never have a perfect force for the next fight, the best hope is that the force you have can morph into something that meets the needs when the time arises. An adjunct complication now arises. Fifty years of virtual peace for the West has meant that most Western militaries have prioritised utilisation over utility. That is to say that priority and resources have been given to those military assets that can be used on a day-to-day basis for security operations rather than those that might be needed for that high intensity combat. The only exception to this set of norms has been the rearmament of the US and its focus on retaining combat edge against all potential adversaries.

Similarly repeatable is that this same balanced force across air, land and sea capabilities has not been (and is unlikely to be) overturned by the introduction of cyber, information manoeuvres or developments in space domains. Such capabilities are instead an *addition* to regular battlecraft in the same way, perhaps, that tanks and submarines and aircraft have been regular warfighting assets over recent military history. Indeed, this primer's evidence generally weighs against the new capabilities of these fourth and fifth domains and their revolutionary effect that many commentators report, the upshot here being that failure to integrate these capabilities may be significant but likely not disastrous to those parties' force design. As throughout this primer, a verso exists (indeed, versos have been a useful device throughout the book, if only to illustrate how frictions exist across most norms and provide general inertia to kneejerk change in those norms). In this case, it will be failure to invest and operate in the widest field of capabilities that will provide an adversary with advantage in future conflict.

There would appear to be similar hubris in Western professional military education (which has in turn trickled into passing political dialogue). In some undefinable sense, military professionalism has evolved such that Western parties now rely upon defeating adversaries by combining excellent traditions and a better 'determination' to fight with the presumed competence of Western militaries. It is this recipe that will assure victory. Although this may appear annoyingly reductive analysis, the trend it suggests looks like groupthink with, this primer observes, a diminishing basis in evidence. An example is useful. While technology can make war more efficient, if both competing sides have that technology, even a highly efficient war is likely (of course) to involve the same enormous costs in blood, metal and treasure that have characterised previous wars. There is no new norm here. But an adjunct observation becomes that armies without the size and depth to absorb losses and remain viable on the battlefield 'may find that no amount of digital wizardry or tactical nous can save them'.[4]

---

[4] Economist Editorial, 'Ypres with AI', *Economist Special Report on Warfare After Ukraine*, 8 July 2023.

Indeed, campaigns throughout history reinforce instead the enduring notion that the enemy always gets a vote. It is, moreover, empirically the case that Western intelligence agencies have usually failed to understand the intent of adversaries, their plans, and their ability to learn, to adapt and then shift their own ways of fighting across battlefield practices.

While none of this should be a surprise, the fastidiousness and inflexibility of current Western planning processes takes little account of the activities, scope and then the ambitions of the enemy. This must impact norms, the more so given that the argument that adversaries are unlikely to abide by the same behaviours, constraints and conduct on the battlefield as those practised by the West, a facet which is then characterised as 'strategic surprise' in an a posteriori justification for the failure of Western militaries to deliver on Western politicians' demands. Here, it is norm *divergence* that again explains a deal of difference in how adversaries are likely to fight over the coming decades, be that in the tactics they use, the ethics they apply, the morals with which their forces are imbued or the empirical realities of how they fight. In each of these cases, what is exhibited on the battlefield will certainly depart from those norms and expectations that Western military have come to assume (and, the authors note, that are set out in this primer). It is therefore the assumptions and context that explain these differences that will remain pivotal to the unpicking and then understanding of these norms.

Nevertheless, Western military activities, societal attitudes and general expectations around war have continued to evolve (and to do this at breakneck speed) since the end of the Cold War, taking on the attitudes and moral behaviours of the societies they represent. The Western mindset towards fighting, to death and destruction, and also to collateral damage, all of the components of war's fundamental nature, have changed markedly since the 1970s. Here, attitudes towards antipersonnel mines, to chemical and biological weapons, to civilian casualties, even to waste and available efficiencies have all developed considerably from the institutional straitjacket that had characterised democracies' earlier battlecraft. But not all parties have experienced this same 'maturing' that has been undertaken by the West. Whether in Russia, Iran, China, North Korea or their proxies around the world, those adversaries show much less concern for Western sensitivities. Their decision processes are systemically opportunistic, transactional and dominated by real politik and this will continue to explain a large part of the overt divergence in normative behaviours between East and West, between autocracy and democracy, peer and near-peer. Indeed, the changes observed in erstwhile stabilisation boundaries mean that it is likely more and more difficult for politicians and leaders to address key concerns in a language that long-dated and polarised adversaries can understand and trust. Moreover, very similar differences exist in the basic currency of *political* discourse between these international blocs, making disagreements harder to mediate and, ultimately, wars more difficult to end in any reasonable way for those involved parties.

A further conclusion may be more interesting for norm definition. Warring parties continue to measure war's costs and gains in very different ways. For the West, the calculation and articulation of war's costs depends on how the reporting party is being measured and depends generally upon audiences and the agenda that decision-makers are seeking to have adopted. It is also about the construction and dissemination of narratives. The US-led occupation of Afghanistan, for instance, later moved into a nation-building mission and is reputed to have cost more than US$5 trillion. By comparison, Russian assistance to Syrian leader Bashar Al-Assad cost a fraction and yet delivered the assurance to the Kremlin of a Russian proxy state on its southern border with key influence in the Caucasus, in Syria, and across the Middle East more widely. Whilst American costs were measured in blood and treasure and in overt calculi made by both its public and political classes, Russia's interpretation of costs continues to be judged exclusively in terms of influence gained. Autocracies, moreover, have more levers (and also less need) to win the narrative. And, while it is hard to judge the relative merits of how these wars are measured at specific points of history, it might seem at this early juncture that Russia's more salutary interpretations have created greater strategic clarity and, indeed, an end state with which it will likely be happier.

It is also possible to attribute some of this imbalance to the idea of technological determinism which has taken such strong hold of the military and, more recently, of the political classes in the West. The idea, after all, that technological superiority determines the outcome of military campaigns has been a foundational Western belief since even before the end of the Cold War. Yet there is a lack of evidence to support these claims, notwithstanding that they still form a core assumption to Western military planning processes. As highlighted by France in his book *Perilous Glory*, technological advances tend to be fleeting on the battlefield and they rarely pay the dividends that is so often expected and promised by them.[5] For Western militaries, therefore, to invest so much capital (intellectual and financial) in achieving overmatch in

---

[5] John France, *Perilous Glory: The Rise of Western Military Power*, Yale University Press, 2011.

technology (almost to the exclusion of other forms and means) as a decisive competitive edge appears ever more paradoxical, the more so given its poor history in delivering clearly attributable success on the battlefield. Instead, outcomes (and, by extension, norms) are better informed by superior tactics, better fighting and morale, better economic heft as well as a more creative cast of military leaders all the way across the spectrum from non-commissioned officers to senior commanders. Versos of course exist but here are more confusing: technology may continue to confound adversaries but a quite separate norm remains that lesser technological solutions can be tactically decisive and, indeed, munitions that might not be at the cutting edge of technology can still offer significant operational advantage on the battlefield. Take, for example, the use of long-range precision munitions in Ukrainian counterattacks against Russian forces. Here, Ukrainian general staff have been able to employ systems such as HIMARs to attack deep in the Russian logistics chain that has taken account of the way the Ukrainians re-supply and fight. By attacking rail infrastructure, command and control nodes, fuel dumps, and critical supply nodes, Ukrainian forces have been able to shape and influence the way that Russia has been able to fight. But an enemy is rarely dumb. Dispersion of assets and quick introduction of countermeasures tend to level briefly unlevel playing fields. Similarly (and as reported above), Russia was able to tie down and fix the bulk of Ukrainian first division forces earlier in its campaign in the Donbas region with mass artillery fire in a Soviet doctrine derived from experiences on the eastern front between 1940 and 1944.

The idea, also, that precision-guided munitions have alone provided more operational success than aged artillery pieces is also less than convincing once emotion and Western interpretations for a Ukrainian victory are removed from the equation. After all, several ingredients are required for sustained military success (and failure) on the battlefield and it remains moot whether Western militaries are learning more widely from very recent campaigns and making conclusions which are grounded in current practice, current playbooks and wishful thinking. That capabilities based upon lesser technology can be effective on a battlefield still appears at odds with the belief set that the West's military leaders, its politicians and defence industries continue to argue. To this end, Ukrainian practice confirms that older capability assets (be they begged, bartered, recovered or refurbished) generally continue to have huge utility in today's high intensity combat.

Appropriate tools and study, however, remain the principal means available to help parties make valid conclusions from contemporary campaigns (the useful notion, borrowed from machine learning, of one experience being transferrable to the next, albeit slightly new scenario). This is then mediated, of course, by the long course of each party's societal change, its experiences of war, warfare and combat. The same is true for norms. Politicians, planners and commanders must pay similar attention to the passing norms that are in front of them. Most balanced forces, after all, can perform across multiple scenarios whether that be the urban fighting in Syria and Iraq, the manoeuvre warfare being employed during counterattacks against Russian forces in Ukraine, or the positional warfare being executed in Yemen. The takeaway here is that change is generally and empirically incremental. And, while the expectation remains (not the norm, of course) that militaries will fight better with that new piece of equipment, the reality remains that militaries eventually adapt well and innovate with the equipment with which they have to fight. The trend is even more pronounced given that few Western militaries have retained legacy equipment with which they can kit relief forces, reserve forces, militia forces or other proxies raised from wider society. As in Ukraine, militaries are not able to pick the weapons with which they will fight once their first line stocks have been depleted. Priorities must then adapt with the fight being based instead upon equipment that is available either off-the-shelf or from generous parties not yet engaged in that combat operation. An adjunct consideration then becomes whether Western militaries can retain sufficient skills and capacity when so much of their training has been specific to high-technology equipment.

Linked to this is a further hypothesis that should at least query the importance afforded to precision as a military asset. In much of the combat currently being undertaken on the world's stage, 'good enough' solutions as well as fixes that involve force multiplication (and other abundant low-level assets) are usually more than satisfactory to achieve political ends. This may not be a popular conclusion for Western military procurement or industry suppliers that have decided that technological superiority is the prerequisite for success. Nonetheless, an acceptance of 'good enough' versus 'boutique' and 'exquisite' equipment is likely to alter significantly the cost of war fighting over the period of interest to this primer. This is particularly the case in subsequent stages of a conflict when stores and war stocks are suddenly less abundant.

Given, then, a requirement to operate and maintain a variety of equipment with which frontline militaries may be unfamiliar, a norm is reinforced around the prioritisation of engineers and machinists who are able to adapt and maximise available equipment. The same priority extends generally across skilled personnel (such as medics) but stoking tension as states wrestle with retaining these skills within an otherwise civilian workforce against deploying them on the front line with regular troops in order to match combat needs. Current practice (and therefore norms) suggests that retaining those personnel in-house on the military's balance sheet amplifies the wider ability of a party's available troop cohort, allowing the personnel to adapt and learn better. In modern military parlance, this bolsters parties' ability to deliver *adaption* across required battlefield practices and where, by extension, early investment in the intellectual and professional development of junior ranks and ratings will continue to reap reward in future deployment (the more so, of course, given increasing sophistication and specialisation across means and forms).

Here, empowerment of junior ranks becomes increasingly important as means and practices develop across tomorrow's battlefield (but also given the fog of war and the need for connected leadership across future domains of war). The primer notes again and again that war is not getting any easier. Instead, the period under consideration will see conflict get quicker, sneakier and more layered and will involve ever more attack angles. Western militaries have spent too much time in quiet denial around adversaries' foundational capabilities, for instance the utility and ubiquity of hostile electronic warfare. The spread, pervasiveness and relative low cost of these competencies (as well as their broad democratisation) have made a myth of being able to operate a seamlessly connected force. Western military experience is still rooted in recent campaigns where effective jamming and electronic blackouts has been quite limited (and much inferior to that both practised by those non-peer adversaries elsewhere in their own conflict zones). Electronic warfare tools are widely available regardless of peer/near-peer status. Not without reason, combatants in Nagorno-Karabakh developed a saying that 'transmitting on the battlefields spells certain death'. The reality of ubiquitous electronic surveillance means that radio transmissions are identified and targeted within seconds.

Technology notwithstanding, it also feels very unlikely that the fog of war has been overcome. An unhealthy relationship emerges between manufacturers' promise of data-driven clarity, commanders' increased reliance upon these purported tools, changes in behaviour resulting from their adoption and then ensuing challenges as intermittent connectivity and adversarial work-rounds compromise blue team performance. And just as electronic warfare hobbles the idea of a newly connected force, resurgence in the adversary's use of deception, secrecy and misinformation (as well as the challenges arising from incorrect or misunderstood data feeds and the like) will likely return renewed emphasis on old paradigms, the role and efficacy of the operational planner, on operational art, informed delegation and empowerment as well as the critical role of the commander. Indeed, Western military forces have traditionally performed well in terms of command and control notwithstanding concerns around the recent growth in scale of headquarters as forces try to keep ahead of the battlefield's increasing complexity.

Each one these conditions feeds to a varying degree into the forms and norms of current conflict. And, while several of the primer's conclusions may rankle readers, comfort is always to hand given the raft of barriers and other sources of inertia that persist to dull the pace of change. Some of these hindrances are structural but, as discussed throughout the book, most are behavioural and contextual. Still others have their foundation in the series of big strategic concepts which the West has continued to bet upon since the end of World War II. This, however, is to ignore that that certain of these same big ideas (even those of deterrence, containment and coercion) have demonstrably not been as successful as many commentators conclude. It is this disconnect that calls for all of these principles to receive fresh analysis and this primer's attention.

In all of this, one salutary lesson stands out ahead of all other salutary lessons. It is always *context* that is the key determinant. In understanding norms, context overrides almost everything else. Each campaign, after all, is markedly different from its peers, whether in terms of the adversary (and how it fights), the permissiveness of the environment, the entrenched nature of quarrel over which fighting occurs, the relative merits of each force including, inter alia, issues of personnel, logistics support and training, and, critically, their commanders. Differences may then be occasioned by levels of preparedness, by the contours of each protagonist's narrative, by luck and, ultimately, by how outcomes are engineered and played. Winners, after all, have the upper hand in moulding both the narratives and norms that occupy this analysis. And just as a party's 'will to fight' is the pivotal piece here, it is still shaped by political leaders and the role of each party's society in shaping the fight.

It is really at this point that the primer's analysis chips in, both in how to consider passing norms and in how best to factor for parties' narratives and priorities over what is likely to be a prolonged campaign. It is for this reason that the reader was encouraged at the book's start to consider specific norms within the guardrails of the individual chapter headings and to situate trends within their context. Readers can then consider the matter within the framework of the *whole* book in order to think about the tapestry of norms, conventions and expected behaviours (and, of course, the drivers that are likely to move them) as they might present over the coming two decades. It is, after all, the *meld* of these factors that comprise the norms of any particular moment. Yet, unhelpfully, it just does not work to assign each factor its own weight in that mix. Nor is there a nice clear formula that jumps from these pages to help readers to decide upon those loadings. This, of course, mirrors the battlefield where there is similarly no preordained route to victory or, despite the lot of the poor planner, any magic playbook that can be followed which will guarantee success. Nevertheless, shuffling one's pack to reflect all of this and acknowledging the fluidity of norms' underlying equations will, the authors believe, place the odds in that side's favour. Where national interests, resources, ideas, values, beliefs and lives are at stake, decision-makers cannot simply accept a set of current presumptions that is presented to them. Instead, those decisions must calmly be rooted in evidence, and informed by perspective. They must take due account of war's behavioural components and, finally, be underwritten by the insight that context is everything.

Appendix I:
Executive Summary from Primary Evidence (non-attributed)

Context's Continuum

Norms

- Not without fault, today's international system behaves as a rules-based arrangement in which bodies such as the United Nations provide an outlet for disputes among nations;
- This arrangement provides ballast to previous balance-of-power systems which relied on brute force and coercion to solve problems within the international system;
- A relatively stable set of rules still exists for state behaviour in conflict and warfare notwithstanding ambiguous and indeterminate terms. Note: law can thus present a resource that states use to justify their practices;
- Norms are about patterns of behaviour as matter of fact but also about expectations of behaviour (moral expectations informed by principles);
- Also, norms are not just legal frameworks (rules of war, often set at lowest common denominator and allowing considerable latitude). Behaviours are actually much more restrictive and mostly self-imposed. An extension of natural culture;
- Thus, a principle of action is binding upon members of a group;
- Such norms are actually quite stable and historically enduring and, to the extent that they change, they tend to become more (not less) stringent;

- Even when there are occasional violations, self-correction tends to happen;
- Norm adherence is thought to have an attractive stabilising influence on behaviour;
- Norm violation versus norm degradation is a key question;
- All norms are contextual and differ by geo-politics or region;
- Lack of congruence among strategic actors in an increasingly multipolar world further deters interstate war. Nevertheless, this does not prevent war but instead increases the frequency of small wars; ensuing conflict will therefore be constant, limited and indirect;
- Strategic actors, inspired by cynical policy objectives, will use the international system/rules/processes to compete in the leaden space that borders war and peace;
- Strategic actors will use the international system in order to compete so as to obfuscate their involvement and attribution, and to prevent that involvement from driving the conflict into large-scale war between other great or regional powers;
- State silence can also contribute to the 'acceptance' of new norms by default, which is problematic because such new legal norms are not international in scope (based on the practices of a very limited number of states);
- How states use novel weapons leads to the emergence of norms. These are not legal norms, but social norms, centred on understandings of appropriateness that are often implicit rather than explicit.

Different Strategic Context?
- Western way of war is the American way of war. Everyone else is playing catch up. The American way of war is also accelerating faster than anyone can establish a baseline. Healthily led by ideas not by technology;
- But the notion of the American way of war based on 'World War II with better technology'. The new norm is that there are many ways to win war;
- Strategic context depends upon an observer's set of assumptions. Thus, revolution here is in its observation. For instance, conflation between disruption in personal tech (iPhone, Internet of Things et al.) and carrying across those expectations to battlefield disruptions. Decade-long trends have become more apparent;
- Vast numbers of small adaptions can add up to very quick and fundamental changes in wartime as a direct response to adversaries' actions (versus slow-moving, top-down incrementalism of peacetime);

However:

- Continuing buy-in of conventional thought despite 'digital dreadnought' moment?
- Drag of doctrine (versus quick changes);
- Current focus is too much on capability rather than intent. Note: threat = capability + intent. Capability is much more distributed with democratised weapons and proxies. More important is political will in its broadest sense (the will to fight). No change here in history. Political will trumps all but has recently been drowned out by the clamour around capability;
- If you cannot out-buy the enemy's political appetite; you must accept that you are going to take the first hit. In this case, resilience is key (military and societal = combined issue);
- Defence now specialised to such an extent that responsibility to defending the people has been forgotten, for example, the state's people and not the state itself;
- Eroded requirement for worst-consequences' planning;
- Law of Armed Conflict (LOAC) outpaced by technological developments;
- New norms usually require a lot of blood. Now we have a bloodless conflict, so where do new norms then come from?

Discontinuities?

- Widening definition of war (Kilcullen's 'conceptual envelopment'); blurring definitions risk misunderstanding and/or complacency; 'vertical escalation' (increasing intensity of action within given location/category of competition) and 'horizontal escalation' (expanding geography/categories/actions); also, 'horizontal manoeuvre' = creating a bandwidth challenge by expanding spectrum of simultaneous small disputes/tasks;
- Material acceleration in capabilities (exploitation of dual-use technology, access and collapsing cost, proliferation) and evolution towards 'pacing threat' (matching of benchmark adversary);
- Rise and rise of good-enough solutions, very-near-peer;
- Quick adjustment/acquisition of capabilities as direct adaption in response to own or others' crisis (lessons from other people's wars');
- Issue of transparency, changes in rear/near;
- Development speed of technology but note: integration issues, inertia agents, digital backbone vulnerabilities;

- Relevance of marginal/incremental meddling versus material/wholesale change? Note: this is a process not an event;
- Real relevance of new tech? Changes to OODA (observe, orient, decide, act) speed, information manoeuvre?
- Trans-military/non-military war options: financial disruption, foreign exchange/stock manipulation, exploitation of humanitarian aid, cyber/information warfare, narcotics trafficking/smuggling, influencing tech/tech standards, lawfare, resource warfare, sanctions, purchase of strategic real estate;
- War fought/won beyond the battlefield;
- Dependency issues around tech, aversion to casualties, need for international support;
- 'Attacking birds with golden bullets': issue of war 'extravagance' and increasing disconnect between expensive weapons and their targets.

Resilience
- Accelerating sophistication and prevalence of misinformation, disinformation and global media;
- Social media, world-with-a-phone; rise of digital authoritarianism;
- Importance of moral authority? Science, homeland resilience, societal mobilisation, new or re-definition of total war, degree of UK denial?
- Unexpected new requirement for different supply models, stockpiles, buffering;
- Decision-making's speed (especially around first contact/rebuttal/first action);
- Relevance of alliances;
- Issue of deployment speed, degree of readiness (general versus cyclical);
- Procurement cycle out of kilter with changes in kit and methods;
- Complexity/crucial issue of UK Strategic Brand (resulting allocation, doctrine, modus operandi);
- Over-promise/unjustified hope in tech, cyber;
- Force multiplication; autonomy and proxies and mercenaries or private military companies (PMCs)?
- Trend towards military action based on information warfare and arm's-length instruments (drones and mercenaries to provide deniability/strategic ambiguity) enabling intervention without the risk of entanglement – synchronised, systematic and strategic;
- Plus, traditional diplomatic instruments of arms control/counter proliferation have disappeared.

Sub-Threshold Activities?
- Compete for strategic advantage without triggering military response;
- Note: West's openness, values versus others' digital authoritarianism and central planning;
- Expanding roles of PMCs, proxies, mercenaries, but not new; new degree of influence?
- Rise of remote arm's-length instruments, information warfare;
- Disappearance of traditional diplomacy/diplomatic instruments, absence of sanction;
- Persistent competition = similar to state's perpetual/background crime activities;
- Societies accept a certain level of such activity but over threshold elicits disproportionately large state response. And are constantly developing. Each society decides its own threshold and reaction.

Deterrence
- Very difficult to measure deterrence and its utility. War is getting sneakier: blurring of three components of capacity, will and signalling;
- Rise again of Maoist strategy of framing victory in terms of adversaries changing their minds?
- UK is in the business of deterrence not defence?
- New deterrence works by changing the adversary's strategic calculus;
- Goal is minimum deterrence that is credible in any particular environment;
- Issue of force deployment's width, depth, sustainability;
- UK is no longer a high-end fighting force: no depth, notion of a 'just short of a war' war?
- Role of other forms of coercion, non-lethal;
- Issue of spiralling costs (hardware and software);
- Need to protect centre of gravity (alliance cohesion) also thematic communities of interests;
- No passivity; remember the five ingredients of deterrence – comprehension, capability, credibility, communication, competition;
- Note: the importance of countering adversaries' fait accompli strategies.

Existential Peril?
- Not to UK homeland but note: adversaries' sub-threshold/low-cost/persistent disorder/competition/sowing instability;
- Adversaries' recent learning phase, targeting/allocating versus UK vulnerabilities;

- Adversaries' improvements to own resilience to absorb strikes, deny West's ability to bring force to bear.

Schizophrenia re Use of Military Instrument?
- Need for ally cohesion, thematic communities of interest;
- Adding competition to existing deterrence mix (together with capability, comprehension, credibility, communications);
- Requirement to mitigate adversaries' fait accompli strategies (unknown equivalents of, say, South China Sea, Ukraine models);
- Note: problem/issue of UK 3-5 year planning cycles so change difficult to predict for longer-term issues;
- What are we resourcing? Priorities, outlook/timeline?
- Fixated on 'tomorrow'? No consideration of roadmap/pathways to outcomes?
- Price of everything, value of nothing?
- Problem of fighting next war with last war's capability/doctrine/kit? Note: procurement remains so long-dated;
- Tokenism; South China Sea visit in 2021?

How Will Militaries Fight Wars?

Mass Times Acceleration
- Force always remains mass times acceleration. Has tech altered this equation? If so, how? Question whether UK paying for tech at expense of mass;
- Smaller and faster to avoid detection, stealth, passive deception, cheaper munitions/assets, integrated, open architecture for new capabilities? But relevance of this over longer term?

Warfighting?
- Decline in experience base, poor response to recent poor performances, subsequent loss of institutional knowledge;
- Issues of logistics (readiness, transport of not-in-theatre assets, replenishment and sustainability, deployment speed);
- Premium on adaptability;
- Greater dispersion leading to greater empowerment of juniors;
- Unpredictable timeline/duration after first contact; note: importance of Day One capability and rehearsal;
- Expense/ramifications of UK soft issues/changeovers/health and safety (H&S)/range and ops/training restrictions;

- No rear, near, close but 'in the people'?
- Issue and untested ramifications of Western casualty aversion (note: contributory reason behind remote warfare);
- Casualty aversion exists primarily in minds of military and politicians, not in public?
- Human forces kept remote, preference for stand-off engagement;
- Promise of technology or unrealistic 'revolution in expectation'?
- Casualties also includes hardware. Materiel and 'unlosable' exquisite platforms?
- War versus non-war (note: important different state legal/governance apparatus: war – permissive/LOAC, non-war – restricted);
- Role of information manoeuvre;
- What is now meant by 'imminent' and 'force'? Trend to other coercion, exacerbated by covert practices?
- Use of Explosives in Populated Areas (EWIPA) initiative; ramifications of public pressures versus the practice and effect on ops/doctrine;
- Tipping point of bomb saturation rather than smart/non-smart munitions mix as key driver in EWIPA argument; note: emerging consensus on mitigation.

Issues of Legitimacy, Morale, Buy-in, Engagement
- UK at key advantage in capability of fighting/public/service people. Cf temporary effect of Novichok/Salisbury event = UK victory/change in Russian practices?
- Buy-in = permission;
- Note: difficulty of changing organisation where previous practices have worked;
- Issue of mission confidence rather than task confidence.

Adaptability
- Issue of procurement cycles still at 5-15 years+;
- Importance of adaption given next conflict type cannot be predicted;
- Adaptation gap = degree to which parties must adapt is increasing versus quickly. Requires faster means/pathways/priority of adaption. Cf pace of tech change in assets that have military use (hypersonic, uncrewed equipment);
- Note: advent of two new domains (cyber and space) versus two domains for most of history (land and sea);

- No ability to generate new weapons in short order therefore requiring increased priority to reorganising, 'fighting current weaponry differently', priority of new combinations, importance of agility and ad hoc, command and control changes to accelerate speed/decisions.

Liminality

- In liminal war, strategic actors test the threshold of acceptability while laundering their involvement;
- Liminal = 'threshold', describes the ambiguity experienced by entities transitioning between two states of being; limbo, on the periphery, ambiguous; note: liminal status of guerrillas, refugees, militias, terrorists, resistance fighters;
- Political/legal/psychological status or whose very existence is debated; liminal geographies = lack of sharp lines;
- Neither fully overt nor clandestine; taking sufficiently few and ambiguous actions to achieve political objectives below military reaction (= 'liminal manoeuvre');
- 'Liminal phenomena' ideally suited for weaker players;
- Surfing the threshold of detectability; sometimes sub-liminal (literally, below threshold of perception) or otherwise in open to seize advantage;
- Exploiting undefined or legally ambiguous spaces or categories as cover for action without retaliation;
- Liminality helped by pervasive electronic surveillance, social media landscape, tightening constraints on democratic governments requiring time/proof before action;
- Strategic actors, both state and non-state actors, use the international community's rules, red lines and thresholds as a handrail for exploitation, despite their intended purpose to constrain self-interested action;
- Liminal war manifests in three practical ways:
  1. strategic and tactical shaping operations through non-lethal capabilities, such as cyber and information operations;
  2. war by way of surrogate, habitually referred to as proxy war; and
  3. fait accompli: Seize a piece of territory, then try to avoid war. Exacts a small gain from an opponent, with sufficient credible force to compel the opponent to accept the loss rather than retaliate in kind;
- Plausible deniability in liminal strategic space; appeal of non-lethal operations;
- Three liminal zones: 1) detection threshold (clandestine actions discovered), 2) attribution threshold, 3) response threshold;

- Liminal activities also at side-show / diversion attributes and possibilities for temporal ambiguity as well as 'reflexive control' (causing targets to act in the interests of the propagandist / actor without realising that they have done so);

How Will Conflict Be Waged?

Changing Domain Mix and Key Questions
- Big current question = 'What is war today?'
- Note: overt combat is tiny component of combat; cf contest versus conflict, small percentage of those under arms at deployed combatants;
- Consider redefinition of conventional, hybrid, asymmetric?
- Near peer advantage arising from disguising war as peace. New threat is therefore not being strategically imaginatively enough;
- Asymmetric war is *not* unusual. Asymmetry always present. All warfare asymmetric (bigger, different capabilities, different tech = asymmetric but not fundamentally different). first key question = 'what is conflict' going forward. Armour or election interference?
- Second key question = what is acceptable today regarding democracies' strategy? (given it must be both feasible and acceptable);
- Key choice is increasing strategic intelligence and engaging with key questions ('what is war, what are our boundaries, moral cersus immoral?);
- Note: people do not feel the threat of, for instance, disinformation or criminalisation (cyber, ransomware) disguised as peace the same way they did as Cold War missiles;
- Post Kuwait a new Western narrative around weapon precision (notwithstanding precision weapons' tiny proportion relative to the use of dumb munitions);
- Attendant danger of casualty-free conflict; note: also, unrealistic military medical effort and casualty evacuation (CASEVAC) promises. Disproportionate emphasis created by national press. Issues of guilt;
- Also, morphing dynamic in rules-based approach of conventional conflict by Western powers because of pervasive media = potential weakness (plus important new priority of creating / owning narratives);
- Ubiquity of information is the only new element / substantive tech change. 24 hours news media. Cascading use of disinformation not new but its global and targeted spread is new;
- Rising importance of 'good enough' technology? Especially, precision capabilities;

- Unexpectedly quick pace of catch-up for near-peer; note: this is not to ignore ongoing role of organised violence and high intensity conflict;
- Competition for strategic advantage without going to war using attacks below the threshold that would prompt a warfighting response;
- Next generation munitions with ever-diminishing returns; disproportionate investment for just marginal advantage;
- Adversaries' improvement of own resilience to absorb strikes, to deny ability to bring our military power to bear;
- Rising vulnerability of supply chains, replenishment, rear area, sustainability, five domains but also financial/soft domains;
- Ever more expensive for US to maintain dominance across expanding domains;
- Requirement thus how still to achieve strategic objectives and counter adversarial narratives (plus issue of what/priority tools for this);
- Dangerous temptation is that military instrument as obvious answer;
- Actually, new key = domestic countermeasures key (population cybersecurity, critical reading/awareness of media, countering notion of outsourced security by tax-paying public). Note: adversaries deliberately avoiding military so military response increasingly pointless?
- Liminality: use of hackers, cyber militias, relationships with organised crime networks, manipulation of migration, alliances with 'useful idiots', development of extensive intelligence, use of active measures/ assassination;
- Scandinavian Total Defence models interesting notwithstanding different context = more blending military and civilian/industry at group responsibility. Useful vehicle even if only to enable these conversations.

Battlefield Asset/Capability Mix

- Debate not about numbers/platforms but about intent and how do we combat this;
- Not just re platforms but also organisation (for example, command and control in persistent engagement?). Note: political pressure to talk about platforms;
- 'Glass cannon build' syndrome: weapons capable of huge effect but disproportionately weak when themselves hit?
- Precise warfare very inefficient (£200,000 per pop versus soft-skinned vehicle?); is this sustainable?
- Military budget erroneously covering cyber and other non-military responsibilities; requires division/clarification;

- Pushing combat into urban areas where new/complex systems/tech will be degraded (see below);
- Issue of mass; role in deterrence?
- Further key = how to organise on battlefield. Dispersed bodies that are plug-and-play to deliver mass to dynamically create right organic formation for task on point;
- Areas of exception exist; cf Chinese shipbuilding and danger of US misunderstanding?
- Dearth of experience versus too heavy reliance on experience? Similarly, political and military expertise gaps?
- Key relationship going forward = the continuum that exists from full time regulars to reservists to veterans;
- Military must also engage with private sector going forward, including local PMCs (note: best local ear and community relationships, best intelligence gathering).

New Sophistication of Covert? Use of Remote Warfare (RW)

- Plausible deniability, accountability;
- Less space for negotiation?
- Controlling airspace below 200 feet? Drone countermeasures?
- Media element, information dissemination and legitimacy challenge;
- Remote warfare = special forces, local protection groups, networks based on future plans/scenarios/retaliations, deterrence (cf joint exercise coordination), visible reinforcement to signal consequences to adversarial actions;
- Note: odd spectrum here between signalling and diplomacy, presence, reinforcement and mix of communicating red lines (rather than defined by capabilities);
- Instead, risk in RW being transferred to local force and local populations;
- Small islands of excellence (cf special forces, elite pockets) possible but lack of knitting into main forces results in reversion;
- RW/local partnering versus near peer threats at trip wire deterrent model rather than alternative of having to mobilise;
- RW deployment vs 1) long-term engagement and 2) to stop an escalation cycle. Also, avoids provocation of having forward bases/deployment;
- Difficulty of term 'proxy' given such organisations have own autonomy/interests/sovereign decision making;
- LOAC cannot be delegated away;
- Continued validity of broad-spectrum capabilities;

- Essentiality of the human;
- Essentiality of political will and 'picking battles';
- Importance of investing time in preparing society for forthcoming decisions and changes;
- Adversaries are the same learning organisations 'that we would wish to be ourselves';
- Note: norms are established in times of competition not conflict, plus gearing society, coalitions, industry, intelligence, diplomacy = whole-of-nation effort to re-establish norms.

Acquisition and Integration of Novel Systems into Legacy Force Design

- Technology certainly not a passport to victory;
- Also requires decisive leader to determine that red line crossed so that such tech platform will be used;
- Tech additive rather than replacing;
- Note: electronics can be fried very easily – thus reshaping of understanding is required. And its susceptibility to electronic warfare;
- Note: also, emanating electronic signature leading to detection on use;
- Language around tech also means actors become dependent upon it = worry/risk for military ops?
- New set of pivotal vulnerabilities; GPS, the World Wide Web, data labelling, rarity of military data training sets, issue of data irrelevance/pollution in training sets;
- Note: potential deployment challenges are considerable, especially embedding of new systems. Here, data storage/handling is particularly poor with no coherent curation processes, made worse by institutional stove-piping hindering any broad revolution in data and autonomy;
- Irrationality of massive projects (cf F-35) that do not go to war (= modern Maginot line). As war becomes more epistemological, winning no longer determined just by battlefield victories – utility of force decreasing;
- Networks and comms architectures unprepared for required data flows, bandwidth and spectrum management;
- Note: operational and doctrinal training/changes required and in tandem; timing/definition/adoption issues;
- General impetus/incentive is for multiple projects at demo/prototype stage but without testing (= 'initial capability push');
- Priority = only at prototype/proof of capability only; not hardened/rested/red teamed. Point of validation and verification is undefined.

Thus, at incremental adoption paths but not wholesale changes. Likely adoption limited to delegation to artificial intelligence/machine learning (AI/ML) in various quite narrow and unsexy tasks;

- Cyber, disinformation – cheap to execute, hard to defend, expensive to surprises and remediate. Reinforces strategic logic that favours David over Goliath;
- Iterative design (design on the move) a disruptor?
- Note: 'side-principal rule' (exhaustion against an adversary's deliberate manipulation to over-commit in a traditional area of strength and conceptual comfort zone); increasingly marginal return on investment?
- Increasingly at age of software importance vs hardware;
- Worrying narrative of leaders' having to engage with technology in order to appear relevant?
- Note: increase in wide untested claims for platform technology: 'F-35 is a great surveillance tool'? Result is 'You don't understand full capabilities so we can mute dissenting conversations'. Note: such platforms do not comport to current paradigm/question of 'what is war';
- Issue: Why double down on conventional warfighting when, in the case of US, already at best? Advances in tech are therefore a canard cf luddites can defeat high tech, especially when politics are wrong. But conflict here = rise of hybrid also as a result of conventional glove's success at maintaining deterrence;
- Invest instead in the weaknesses: change the way we think. Increase our strategic intelligence/better analysis before we invest in any platforms;
- Certain enduring technical challenges remain: contextual reasoning and concept understanding (ability to abstract from concepts). Required computational power too great. Outputs then becoming new inputs = data dependencies/issue of 'undeclared consumer'. Also, interface interoperability issues between legacy and new kit and new plethora of data points;
- Fracturing of disciplines means very little holistic analysis of new tech and whole systems, exacerbated by institutional stove-piping (focus on cost, time, delivery on programme, policy); overly specialised cells in charge of just one slice of system in what is a complex adaptive arrangement with emergent properties;
- Note: further conflict of interest re new exquisite technologies developed in private sector/dual use; tech here might not be offered to militaries for a notional profit margin (as with British single source regulations). Government, and specifically, defence procurement is not

configured for such harsh commercial realities. Also, developers know that if technologies are transferred to militaries, outright ownership of intellectual property and commercial freedoms of action are diluted;

- Government therefore has to be content with paying 'for lessons being learnt', albeit sometimes painful/costly, rather than paying for technological certainty. So, tomorrow's procurement portfolio might contain more incubating programmes than classic, linear equipment-centric projects. A whole new taxonomy of defence acquisition might have to be developed.
- Note: also, current weapons AI work being undertaken with very small number of class characteristics (less than 30). Current challenge is lack of real-world data sets and inability for subsequent sufficient back-test on real world data. Thus, operational convergence/appropriateness remains untested.

Autonomy and Thresholds of Supervision in Lethal Engagement

- Levelling effect of technology, incentive to move to robot/remote warfare systems; force multiplication, casualty aversion (and this is a question), staffing shortfalls, push from private sector, revolution in expectation, suits politicians, cost of human resources (benefits, sleep patterns, CASEVAC/care);
- Nullification and fleeting advantage;eElectronics can be fried very easily – thus reshaping of understanding is required. And its susceptibility to electronic warfare. Note: electronic signature emanating leading to detection;
- Note: number of years since truly significant advances in the fields of AI or ML;
- More ambitious projects that involve machines interacting with the 'real world' remain in infancy;
- Also, infancy/lack of working precedence of ML/AI techniques for safety-critical applications (for example, where human beings may be harmed by the system);
- Safety critical systems have traditionally been built using precise specifications/thorough testing. ML/AI systems, by contrast, are meant to respond to inputs that cannot be fully specified in advance = an inherent difficulty;
- Humans are good at modifying their behaviour to be sensible in different contexts; software actors are not, either because they sense their

environment incorrectly or not at all. Humans then surprised because software is not sensitive to circumstances;

- Thus at 'Potemkin village' with narratives being hyped by tech industry leaders with a financial interest in procurement?
- Vulnerability of precision systems to inaccurate intelligence, their dependence on data and connectivity, and their irrelevance versus amorphous, cell-based enemy; enduring need to translate precision attack into strategic advantage.

Artificial Intelligence and Machine Learning in Military Affairs: Enduring Challenges

- Unhelpful convergence of US tech and technophilic Department of Defense/lobby?
- Under-reporting of systemic vulnerabilities of data-handling that is fundamental to AI and ML;
- Change anything, change everything (CACE phenomenon = 'wicked' conflict);
- Labelling issues (mass, heuristics/biases, 'ground truth' models);
- Issue of data dependencies;
- Suppression of doubt;
- Inference of causes;
- Narrowing of choices;
- Categorising of facts;
- Conditionalising;
- Least variance analysis;
- Issues of data smoothing;
- Coding for ambiguity/vales/aims
- Issue of data forgetting,
- Issues of verification and validation;
- Issue of anchoring/incremental adjustment of representations;
- Issues of confidentiality, integration, user' expectation;
- Unexamined premises linking AI to national security;
- Situational awareness and AI?
- Issue (ethical, technical, operational) of target identification;
- Issue of imminent threat?
- Issue of feint and deception;
- Even if it were possible to create relevant training datasets, changes in human behaviour over time make the initial set of training data invalid. This is particularly true in conflict situations where people change

their behaviour in response to conflict, and efforts will be made by combatants to evade detection. It is simply not a problem that is tractable to automation;

- Data capture and integration challenge given: several languages, five domain integration, sensor multiplicity; what are expected outputs here and who will receive this output? Requires embracing outsource to corporates/contractors;
- Today, AI/ML restricted to routines enhancing human efficiency and efficacy. Toolbox for humans. Humans required in loop, not feasible for being replaced by agents. Also, no answer to adversarial meddling of data (except dynamic analysis of data history);
- Note: issue of humans becoming reliant on algorithm and then doubly prey to spoofing. Need to be aware of limitations of machines.

Battlespace Fighting

Technology Changes
- No transparency, everything is observable but note: hay-in-a-haystack notion, ability to 'move in seams', abundance of noise;
- Electromagnetic spectrum and contested electromagnetic spectrum (note: limited training off air/grid);
- Sensors; sensor density to provide situational awareness?
- Rise of passive sensor but note: in-theatre confirmation required by stand-in sensors?
- Vulnerability to spoof/decoy; 'some sensors, some of the time by some shooters'?
- Precision fires; by 2035, conflict by drones or projectiles?
- Note: tendency to extrapolate forward without context of steadily evolving set of other capabilities;
- Future battle: constant, regionally contained, fought indirectly through proxies, and relegated to urban terrain'
- Role/balance between lethality and non-lethal? Command and control of this decision, impact analysis of this decision?
- Speed of deployment?
- Widespread adversarial 'hugging' of Western tech systems (Google Earth, GPS, hobby drones, iPads) and difficulty of governments' closing down these systems without hampering own operations/publics;
- Role of re-supply/sustainability; munition stockpile versus attrition of decoy/false positives.

Rear

- Increasingly difficult to isolate rear areas;
- Massive expansion in size of operational areas (cf airpower spread more thinly);
- Increased use of bubbles-of-concentration for protection and force focus;
- Disruptive development of passive sensors;
- Logistics/footprint/digital signatures leading to new levels of vulnerability.

Urban

- Cover; consequences of 'infesting' urban terrain;
- Certain durability to infrastructure that is difficult to target in urban areas;
- Levels the playing field re situational awareness given sensors less effective in complex environments;
- Expansion of ever more restrictive legal and political constraints?
- Punishes size, troop concentrations, active comms, hierarchical organisations, overtness, anything with sustained contrast;
- Note: erroneous fighting force density equations/assumptions;
- Defender advantage; difficulty of bringing force to bear;
- Problem remains of defending ground, preventing reinfiltration;
- Mass still required in urban conflict for cordon, screening;
- Weaker adversaries will therefore seek sanctuary in cities to increase their odds of survival and victory.

Velocity

- All echelons engaged simultaneously and on Day One; but actually, little material change in pace since 1615!
- No evidence for leaders' claiming exponential change plus foundation for crucial unfounded/unchallenged assumptions/assertions;
- Day One engagement very fast but slows thereafter as no one has logistics to keep it going;
- Creating a division takes ten years (training, doctrine, experience growth, stores/kit timeline = relevant 'strategic warning time');
- Key issue here = acceleration. Difference = how fast is the movement from first contact to establishing military ops, from identifying problem to establishing political will and deploying military forces. Thus, speed of being able to make decisions and changing/reflecting right character of conflict and exact operation being faced.

Adaption

- Combination of traits conferring survival advantage to those adopting certain characteristics: stealth ('adaptive colouration' and ability to disappear when threatened), dispersion and ability to concentrate for specific ops, modularity and self-healing networks, autonomy and restriction of electronic signatures, hugging/piggy-backing on widespread tech, media manipulation/provocation and exploitation of subsequent errors, political warfare and mobilisation, repurposing of consumer systems/devices to combat systems.

Space Domain

- Critical, contested, congested, competed. Note: space support of air domain, especially commercial airspace. Provision/receipt of PNT signals (position, navigation, *timing*). Legal space legislation at 1967 Outer Space Treaty, no longer fit for purpose;
- Innovation: Rise of nanosatellites capable of surviving radiation/ temperatures; 3,500 currently registered working satellites, PNT now deliverable from low orbit;
- Kessler effect of collisional cascading and consequence of unusable orbital plane;
- Rise of RPO (rendezvous proximity operations): manoeuvre into line of sight, data steal. Soft kill mechanisms, depleting asset life through collision avoidance;
- deterrence ramification from destabilising in-orbit based capabilities.

Electronic Attack and Defence

Cyber

- Lead-times?
- Very specific to target;
- Recce/knowledge exercise, en masse, low probability of success;
- Crafting payload without being detected
- Administrative privileges, multi-factor authorisation;
- Obsolescence, imprecise triggering and outputs, difficult synchronising, likely temporary effect;
- High classification, fuzzy authorisation, difficult dissemination/ allocation;
- Multi-agency op; multiplicity of friction points, difficult to synchronise with kinetic or other action, multiple redundancy required;

- Commercial vs hardened military software platforms? Note: new online presences = expansion of attack surfaces, especially in infrastructure;
- Lack of doctrine here is material issue; matters of sovereignty, escalation, panic, disproportionate means/ends, plausible deniability, actioned in pre-war phase?
- Unintended consequences?
- Actual expert voices get drowned out, signal to noise ratio is awful. Plus, issue of secrecy.

Leadership and People
- Autonomy; ethical issue of deploying soldiers versus new/autonomous machines' sophistication (can we afford not to? What happens when both available?);
- Engagement, buy-in, public support, hearts and minds;
- Morale; buy-in, permission, performance;
- Modern malaises; range time, duration, H&S, ammo, training restrictions;
- Western paradigms at likely inappropriate benchmark; misunderstanding, timeline differences/length of outlook;
- Notion of 'long screwdriver' and technology double edge sword re effect on control on ground. Smallest tactical action can have implications for Whitehall. Tight control measures suddenly 'possible' through technology and its effect on trusting subordinates?
- Conundrum/tension of mission command vs command independence in field? Reflex response is to relegate mission command?
- British army anti-intellectualism still influential. Shaped by society and shaped by civilian norms. Not therefore just a question of military ecosphere but requires persistent and timely management/attention;
- Imprecise effect of 'failure of experts' in recent conflicts ('dodgy dossiers' and the like);
- Enduring requirement for cognitive bandwidth versus 'task-saturation' and focus on the immediate; notion of 'conceptual envelopment' (horizontal expansion of warfare and decisive 'shaping' that stymies timely response to adversarial action); adversarial pursuit, therefore, of conventional *and* non-conventional paths.

Fluid Operational Design and the Future of Tested Campaigning Structures
- Study of warfare is broken or undertaken with insufficient rigour to inform policy. Warfighting analysis now a lonely trade – few interested in

deep analysis of application, logistics. But key here is that commentators cannot opine on future warfare unless they are across current warfare (= capabilities and shortfalls).

- Decision process about how armies are going to fight are increasingly less transparent and less understandable. Note: at applications level, defence reviews are more about equipment and not at all about how to train/fight/organise;
- Dilemma is equation between think tank/academia environment versus familiarity/experience/raw empirics;
- Central difficulty of vacuum between experience versus paper doctrine versus scholarship in order to create meaningful insights. Note: no evidence arises from war gaming, only insights into what might be right questions to ask;
- What makes an army useful? Rapid deployment, being able to sustain at range, being highly lethal, mastering the 'how' of fighting (study/ understanding of logistics, movement, supply, engineering needs to come back to fore) and operating in super-dispersed fashion. Also, importance of killing and capture;
- Other tasks (humanitarianism etc.) only dilute given limited/set number of training days per year;
- Do not mix 'complexity' and plain 'unfamiliarity';
- Conducting mobile campaigns over large areas at long distance requires precise understanding of challenges of movement, logistics, speed, acceptable comprises, training days/costs;
- Pandora box of terms (norms, future characteristics). International relations and security studies use fast-moving concepts which tend to raise historians' suspicions;
- UK malaise perhaps at inaccurate use of 'thresholds' concept'? Cf George F Kennan 1948 *Political Warfare* – UK long-dated mechanism to supplement size of army relative to ground or mission being controlled?
- Analogy of ladder is better than binary threshold? A state can be anywhere on a ladder but can be understood as part of one whole continuum;
- Notion of WaaS – Warfare as a Service;
- What is new here is perception; perception here is UK military's view that thresholds et al. = new. Not true;
- Challenge of manpower reduction = inability to backfill in hot war/ casualties. Also true in materiel and industrial capacity. Need for resilience/buffer/training programme duration;

- UK issue = especially materiel. Logistic pinch point, plus civilian vulnerability to cyber/disruption;
- High readiness is the measure of an army;
- Today's H&S, range safety, ammo constraints, drivers' hours limitations, willingness to accept training accidents;
- UK gets insufficient lethality from every defence dollar spent?
- Un-war creates ambiguity, delay;
- Resetting of customary law.

The Place and Function of LOAC, Human Rights and Humanitarian Law

- Issue of conscience: humans versus robots, humans if robots available?
- Ban on autonomous targeting of people;
- Note: issue that law does not differentiate between offence and defence (attack is an attack, arguments re defence cannot be calibrated);
- Static versus mobile platform differentiation?
- Role of stigma (note: media access, legitimacy); cf EWIPA model: international consensus? Diverging norms on civilian harm mitigation; US mitigation restrictions and obligations; civilian harm mitigation in peer-to-peer conflict?
- Violation of norm versus change of norm; divergence to added stringency?
- Norm adherence pressures: non-governmental organisatinos (NGOs), arms control activists, campaigns;
- Stability of Article 2 and Article 51;
- 'Old laws' more difficult to apply in new tech?
- Norms at contours of political landscape; military characteristic of narrowing;
- Note: limited evidence available from battlefield practices (apart from prototypes/demos) hence rise/need for anticipatory legal declarations of what it might look like;
- Absence of practice means de facto norm changes stymied?
- Also, autonomy is a behaviour, *not* an effect. Hard already to write technical requirements let alone determine what to regulate/prohibit when focusing on warfare behaviours;
- Complexity/conflict in LOAC and overarching importance of human, however dumb/data-starved that human might be. Strange intuition;
- Knowledge gap in NGOs about weapon ops currently in play.

Appendix II: Participants

- Context's continuum; traditionalists, pragmatists and futurists
  - Professor Peter Roberts, Royal United Service Institute (RUSI) Military Science
  - General Sir Nick Carter, Chief of the Defence Staff, Buckingham HRI conference
  - Professor Julian Lindley-French, Institute of Statecraft, Future of Europe
  - General Sir Richard Barrons, ex-Commander Joint Forces Command
  - Dr Sean McFate, Georgetown University, Senior Fellow Atlantic Council

- How will militaries fight wars? Realities, ethics and other empirics
  - Major General Patrick Cordingley, ex Commander 9th Brigade
  - Air Vice Marshall Mike Harwood, ex-head of British Defence Staff
  - Amos Fox, US Armor School, written submission
  - Colonel Dennis Vincent, UK Ministry of Defence HR and Royal Military Academy Sandhurst (RMAS), Centre for Army Leadership (CAL)

- How will conflict be waged? The dynamic of conventional and asymmetric warfare
  - Dr Jack McDonald, Senior Lecturer, War Studies, King's College London

- Ewan Lawson, RUSI and ex-UK Psychological Ops Group
- Paddy Nicoll, RUSI, Advisor 6 Division
- Emily Knowles, Remote Control Project, Oxford Research Group
- Dr Aaron Edwards, University of Leicester and RMAS International Affairs Faculty
- Air Vice Marshall Harvey Smyth, GOC UK Space Command

- Acquisition and integration of novel systems into legacy force design
  - Dr Heather Roff, John Hopkins Applied Physics Laboratory
  - Professor John Louth, RUSI, written submission
  - Dr Nathan Potter, MDBA Emerging Technologies
  - Laura Nolan, Campaign to Stop Killer Robots and International Committee for Robot Arms Control (ICRAC), ex Google, written submission

- Autonomy and thresholds of supervision in lethal targeting
  - Professor Lucy Suchman, ICRAC and Lancaster University
  - Richard Moyes, CEO, Article 36
  - Dr Rob Cole, Palentir Emerging Technologies

- Battlespace fighting: Changes to operations in rear and close quarters
  - Dr Jack Watling, RUSI Land Warfare
  - Brigadier James Cook, UK Army Personnel Centre and Concepts

- Electronic attack and defence; conflict in the electro-magnetic spectrum
  - Justin Bronk, RUSI Air Warfare
  - Tarquin Folliss, Reliance ACSN and ex Ministry of Defence Cyber

- Leadership and people: future command and control structures
  - General Sir David Capewell, ex-Chief of Joint Operations
  - Colonel Miles Hayman, Head Military Strategy, UK Delegation, NATO, written submission

- Fluid operational design
  - Wilf Owen, *Small Wars Journal* and RUSI
  - Professor Matthias Strohn, RMAS CAL and Buckingham
  - Franz Stefan Gady, Fellow, International Institute for Security Studies
  - Major General Mungo Melvin, Enhancing Strategic Capability Study

- The costs of changing norms; proliferation / escalation, deniability / feint, misunderstanding
  - Chris Woods, Founding CEO, Airwars

- The place and function of the Law of Armed Conflict, human rights and humanitarian law
  - Dr Ingvilde Bode, South Copenhagen University and ICRAC, written submission
  - Dr Bonnie Docherty, Human Rights Watch, and professor, Harvard Law School

- The role of enduring cultural and socio-behavioural factors in future conflict
  - Professor William Scott Jackson, Oxford Said Business School
  - Professor Christian Enemark, Ethics Southampton University

Bibliography

A

Adamsky, Dima, *The Culture of Military Innovation*, Stanford University Press, 2010

Allison, George, 'How Much Does a Queen Elizabeth Carrier Cost per Year?', *UK Defence Journal*, 18 July 2021, https://ukdefencejournal.org.uk/how-much-does-a-queen-elizabeth-class-carrier-cost-per-year

Applebaum, Anne, 'Russia's War Against Ukraine Has Turned into Terrorism', *The Atlantic*, 13 July 2022, https://www.theatlantic.com/ideas/archive/2022/07/russia-war-crimes-terrorism-definition/670500

Apps, Peter, 'Black Sea Grain Battle Could Define Ukraine War', *Financial Times*, 30 May 2022

B

Badlam, Justin et al., 'The CHIPS and Science Act: Here's what's in it', McKinsey & Partners, 4 October 2022, https://www.mckinsey.com/industries/public-sector/our-insights/the-chips-and-science-act-heres-whats-in-it

Baer, Dan, 'Six reflections of the first day of Russia's war in Ukraine', Carnegie Endowment for International Peace, 24 February 2022, https://carnegieendowment.org/2022/02/24/six-reflections-on-first-day-of-russia-s-war-in-ukraine-pub-86524

Baillie Gifford website, https://www.bailliegifford.com/en/usa/professional-investor/insights/ic-article/2021-q4-riding-growth-waves

Barnes, Paul, 'Neophilia, presentism and the deleterious consequences for Western military strategy', *Modern War Institute*, 3 June 2019, https://mwi.usma.edu/neophilia-presentism-deleterious-consequences-western-military-strategy/

Beaumont, Peter, and Isobel Koshiw, '"The occupier should never feel safe"; Rise in partisan attacks in Ukraine', *Guardian*, 6 June 2022, https://www.theguardian.com/world/2022/jun/06/ukrainian-partisan-attacks-surge-russia

Bendett, Samuel, 'To robot or not to robot? Past analysis of Russian military robotics and today's war in Ukraine', *War on the Rocks*, 30 June 2022, https://warontherocks.com/2022/06/to-robot-or-not-to-robot-past-analysis-of-russian-military-robotics-and-todays-war-in-ukraine

Bergengruen, Vera, 'How Ukraine Is Crowdsourcing Digital Evidence of War Crimes', *Time Magazine*, 18 April 2022, https://time.com/6166781/ukraine-crowdsourcing-war-crimes/

Berkowitz, Bonnie and Artur Galocha, 'Why the Russian military is bogged down by logistics in Ukraine, *Washington Post*, 30 March 2022, https://www.washingtonpost.com/world/2022/03/30/russia-military-logistics-supply-chain/

Biddle, Stephen, 'Ukraine and the Future of Offensive Maneuver', *War on the Rocks*, 22 November 2022, https://warontherocks.com/2022/11/ukraine-and-the-future-of-offensive-maneuver

Bielieskov, Mykola, 'Ukraine's Territorial Defence Force: The War So Far and Future Prospects', *RUSI Publications*, 11 May 2013, Ukraine's Territorial Defence Forces: The War So Far and Future Prospects | Royal United Services Institute (rusi.org)

Bilal, Arsalan, 'Hybrid Warfare – New Threats, Complexity and Trust as the Antidote', *NATO Review*, 30 November 2021, https://www.nato.int/docu/review/articles/2021/11/30/hybrid-warfare-new-threats-complexity-and-trust-as-the-antidote/index.html

Blum, Gabriella, 'The Paradox of Power: Changing Norms of the Modern Battlefield', *Houston Law Review*, Vol. 56, No. 4, 2019, https://houstonlawreview.org/article/7948-the-paradox-of-power-the-changing-norms-of-the-modern-battlefield

Breaking Defense Staff, 'Vehicle platform integration: Where technologies become capabilities', *Breaking Defense*, 14 October 2019, https://breakingdefense.com/2019/10/vehicle-platform-integration-where-technologies-become-capabilities/

Bremer, Maximilian and Kelly Grieco, 'In denial about denial: Why Ukraine's air success should worry the West', *War on the Rocks*, 15 June 2022, https://warontherocks.com/2022/06/in-denial-about-denial-why-ukraines-air-success-should-worry-the-west

Bruno, Mark, '"Uber for Artillery" – What is Ukraine's GIS Arta System?', *The Moloch*, 24 August 2022, https://themoloch.com/conflict/uber-for-artillery-what-is-ukraines-gis-arta-system

Buehler, Kevin and others, 'War in Ukraine: Twelve Disruptions Changing the World', Mckinsey & Company, 9 May 2022, https://www.mckinsey.com/business-functions/strategy-and-corporate-finance/our-insights/war-in-ukraine-twelve-disruptions-changing-the-world

C

Cappella Zielinski, Rosella and Ryan Grauer, 'Understanding Battlefield Coalitions', *Journal of Strategic Studies*, Vol. 45, No. 2, pp. 177-185, 28 January 2022, https://www.tandfonline.com/doi/pdf/10.1080/01402390.2021.2011231

Chatham House, 'Seven Ways Russia's War on Ukraine has Changed the World', *Chatham House Editorial*, 20 February 2023, https://www.chathamhouse.org/2023/02/seven-ways-russias-war-ukraine-has-changed-world

Chang, Johnathan, and others, 'The Economic Front in Russia's War against Ukraine', *WBUR On Point*, 8 March 2022, podcast, https://www.wbur.org/onpoint/2022/03/08/economic-war-and-russia-ukraine-conflict-sanctions

Cheung, Tai Ming and Thomas Mahnken, 'The Grand Race for Techno-Security Leadership', *War on the Rocks*, 31 August 2022, https://warontherocks.com/2022/08/the-grand-race-for-techno-security-leadership/

Chinn, David and John Dowdy, 'Five Principles to Manage Change in the Military', *McKinsey*, 1 December 2014, https://www.mckinsey.com/industries/public-and-social-sector/our-insights/five-principles-to-manage-change-in-the-military

Clark, Bryan, Dan Patt and Harrison Schramm, 'Mosaic Warfare: Exploiting Artificial Intelligence and Autonomous Systems to Implement Decision-Centric Operations', Center for Strategic and Budgetary Assessments, 2020

Cole, Reyes, 'The Myths of Traditional Warfare: How Peer and Near-Peer Adversaries Plan to Fight Using Irregular Warfare', *Small Wars Journal*, 28 March 2019, https://smallwarsjournal.com/jrnl/art/myths-traditional-warfare-how-our-peer-and-near-peer-adversaries-plan-fight-using

Cordesman, Anthony, 'Learning the Right Lessons from the Afghan War', CSIS, 7 September 2021, https://www.csis.org/analysis/learning-right-lessons-afghan-war

Crane, Conrad, 'Too Fragile to Fight: Could the US Military Withstand a War of Attrition?', *War on the Rocks*, 9 May 2022, https://warontherocks.com/2022/05/too-fragile-to-fight-could-the-u-s-military-withstand-a-war-of-attrition

Cranny-Evans, Sam, 'As small drones shape how we fight, is the British army ready to face them?' Royal United Services Institute, 21 July 2022

D

Davidovic, Jovana, 'What's Wrong with Wanting a "Human in the Loop"?', *War on the Rocks*, 23 June 2022, https://warontherocks.com/2022/06/whats-wrong-with-wanting-a-human-in-the-loop

de Alwis, Akshan, 'A New Age of Multilateralism: Potential Solutions for the South China Sea Conundrum', *Diplomatic Courier*, 7 June 2016

Devanny, Joe and John Gearson (eds), 'The Integrated Review in Context: Defence and Security in Focus', School for Security Studies, King's College London, October 2021, https://www.kcl.ac.uk/warstudies/assets/ir-in-context-defence-and-security-in-focus.pdf

Deveraux, Brennan, 'Loitering Munitions is Ukraine and Beyond', *War on the Rocks*, 22 April 2022, https://warontherocks.com/2022/04/loitering-munitions-in-ukraine-and-beyond

Dubois, Gaston, 'MANPADs in Ukraine: the return of Russian aircraft's biggest fear', *Aviacioneline*, 7 March 2022, https://www.aviacionline.com/2022/03/manpads-in-ukraine-the-return-of-russian-aircrafts-biggest-fear/

E

Economist editorial, 'The Fog of War May Confound Weapons That Think for Themselves', *Economist*, 27 May 2021, https://www.economist.com/science-and-technology/2021/05/26/the-fog-of-war-may-confound-weapons-that-think-for-themselves

Economist editorial, 'The Promise of Open-Source Intelligence', *Economist*, 7 August 2021, https://www.economist.com/leaders/2021/08/07/the-promise-of-open-source-intelligence

Economist editorial, 'The Digital Yuan Offers China a Way to Dodge the Dollar, *Economist*, 5 September 2022, https://www.economist.com/finance-and-economics/2022/09/05/the-digital-yuan-offers-china-a-way-to-dodge-the-dollar?

Economist editorial, 'Globalisation and autocracy are locked together. How much longer?', *Economist*, 19 March 2022, https://www.economist.com/finance-and-economics/2022/03/19/globalisation-and-autocracy-are-locked-together-for-how-much-longer

Economist editorial, 'Has the Ukraine War Killed Off the Ground-Attack Aircraft?', *Economist*, 1 November 2022, https://www.economist.com/the-economist-explains/2022/11/01/has-the-ukraine-war-killed-off-the-ground-attack-aircraft

Economist editorial, 'A New Age of Economic Conflict', *Economist*, 5 March 2022, https://www.economist.com/leaders/a-new-age-of-economic-conflict/21807968

Economist editorial, 'Ukraine's Partisans Are Hitting Russian Soldiers Behind Their Own Lines', *Economist*, 5 June 2022, https://www.economist.com/europe/2022/06/05/ukraines-partisans-are-hitting-russian-soldiers-behind-their-own-lines

Economist editorial, 'What Is Hybrid War? And Is Russia Waging It in Ukraine?', *Economist*, 22 February 2022, https://www.economist.com/the-economist-explains/2022/02/22/what-is-hybrid-war-and-is-russia-waging-it-in-ukraine

Economist editorial, 'Ypres with AI', *Economist Special Report on Warfare After Ukraine*, 8 July 2023

F

Farrell, Henry and Abraham Newman, 'Weaponised Interdependence; How Global Economic Networks Shape State Coercion', *International Security*, Vol. 44, No. 1, 2019

Faulconbridge, Guy, 'Russia warns US against sending more arms to Ukraine', *Reuters*, 25 April 2022, https://www.reuters.com/world/europe/russia-warned-united-states-against-sending-more-arms-ukraine-2022-04-25/

Feldstein, Steven, 'Disentangling the Digital Battlefield: How the Internet has Changed War', *War on the Rocks*, 7 December 2022, https://warontherocks.com/2022/12/disentangling-the-digital-battlefield-how-the-internet-has-changed-war

Financial Times Editorial Board, 'NATO's Weapon Stockpiles Need Urgent Replenishment', *Financial Times*, 31 January 2023, https://www.ft.com/content/55b7ba35-6beb-4775-a97b-4e34d8294438

Ford, Matthew and Andrew Hoskins, *Radical War: Data, Attention and Control in the 21st Century*, Oxford University Press, 2022

France, John, *Perilous Glory: The Rise of Western Military Power*, Yale University Press, 2011.

Friedman, Sir George, 'The Future of War; A History', October 2017, see https://www.youtube.com/watch?v=6xXqrgDP8CA

G

Gave, LV, 'The End of the Unipolar Era', *Gavegal Research*, 11 May 2022

German, Tracy, 'How Will Wars Be Fought in the Future?', *Oxford University Press Blog*, 20 July 2019, https://blog.oup.com/2019/07/how-will-wars-be-fought-in-the-future/

Gray, Colin, *Another Bloody Century*, Phoenix Books, 2005.

H

Halpert, Madeline, 'Russia's invasion has cost Ukraine up to $600 billion, Study Suggests', *Forbes*, 4 May 2022, https://www.forbes.com/sites/madelinehalpert/2022/05/04/russias-invasion-has-cost-ukraine-up-to-600-billion-study-suggests

Hambling, David, 'New Turkish Bayraktar drones still seem to be reaching Ukraine', *Forbes*, 10 May 2022, https://www.forbes.com/sites/davidhambling/2022/05/10/new-turkish-bayraktar-drones-still-seem-to-be-reaching-ukraine/?sh=7c85b64a685b

Hammes, TX, 'The Future of Warfare: Small, many and smart versus few and exquisite', *War on the Rocks*, 16 July 2014, https://warontherocks.com/2014/07/the-future-of-warfare-small-many-smart-vs-few-exquisite/

Hammes, TX, 'The Tactical Defence Becomes Dominant Again, *Joint Force Quarterly*, 103, 14 October 2021, https://www.960cyber.afrc.af.mil/News/Article-Display/Article/2810962/the-tactical-defense-becomes-dominant-again/

Hennigan, WJ, 'Air Force hires civilian drone pilots for combat patrols; critics question legality', *Los Angeles Times*, 27 November 2015

Horton, Alex and others, 'On the Battlefield, Ukraine Uses Soviet Era Weapons Against Russia', *Washington Post*, 20 March 2022, https://www.washingtonpost.com/world/2022/04/29/urkaine-russian-soviet-weapons/

I

ICRC, 'Digital Technologies and War, Vol 102, 913, https://international-review.icrc.org/sites/default/files/reviews-pdf/2021-03/Digital-technologies-and-war-IRRC-No-913.pdf

ICRC, 'Explosive weapons: Civilians in populated areas must be protected', 26 January 2022, https://www.icrc.org/en/document/civilians-protected-against-explosive-weapons

Ignatius, David, 'How the Algorithm Tipped the Balance in Ukraine', *Washington Post*, 19 December 2022, https://www.washingtonpost.com/opinions/2022/12/19/palantir-algorithm-data-ukraine-war/

Inskeep, Steve, 'A big mystery of the war in Ukraine is Russia's failure to gain control of the sky', National Public Radio, 11 May 2022, https://www.npr.org/2022/05/11/1098150747/a-big-mystery-of-the-war-in-ukraine-is-russias-failure-to-gain-control-of-the-sk?t=1654508553446

International Monetary Fund, *World Economic Outlook Database*, April, 2022 Edition, https://www.imf.org/en/Publications/WEO/weo-database

Internet statistics, LiveStats, see https://www.internetlivestats.com/

Ivanov, Zoran, 'Changing the Character of Proxy Warfare and Its Consequences for Geopolitical Relationships', *Security and Defence*, 4, 2020 volume 31, https://securityanddefence.pl/Changing-the-character-of-proxy-warfare-and-its-consequences-for-geopolitical-relationships,130902,0,2.html

J

Jensen, Benjamin and Matthew Strohmeyer, 'The Changing Character of Combined Arms', *War on the Rocks*, 23 May 2022, https://warontherocks.com/2022/05/the-changing-character-of-combined-arms

Johnson, David, 'A modern-day Frederick the Great? The end of short, sharp wars', *War on the Rocks*, 5 July 2022, https://warontherocks.com/2022/07/a-modern-day-frederick-the-great-the-end-of-short-sharp-wars

Johnson, Rebecca, 'Ukraine war shows nuclear deterrence doesn't work. We need disarmament', Open Democracy, 24 March 2022, https://www.opendemocracy.net/en/odr/ukraine-russia-war-putin-nuclear-weapons-disarmament-deterrence/

Jones, Seth, 'Russia's ill-fated invasion of Ukraine: Lessons in Modern Warfare, CSIS, 1 June 2022, https://www.csis.org/analysis/russias-ill-fated-invasion-ukraine-lessons-modern-warfare

K

Kaushal, Sidharth, 'Can Russia Continue to Fight a Long War?', *RUSI Long Read*, 22 August 2022, https://rusi.org/explore-our-research/publications/commentary/can-russia-continue-fight-long-war

Kim, Young Mie, 'New evidence shows how Russia's election interference has got more brazen', *Brennan Center Report*, 5 March 2020, https://www.brennancenter.org/our-work/analysis-opinion/new-evidence-shows-how-russias-election-interference-has-gotten-more

Knapp, Mike, 'The United States is Behind the Curve on Blockchain', *War on the Rocks*, 30 August 2022, https://warontherocks.com/2022/08/the-united-states-is-behind-the-curve-on-blockchain

Kollars, Nina Ann, 'By the Seat of Their Pants: Military Technological Adaptation in War', 2012 Ohio State University PhD text, http://rave.ohiolink.edu/etdc/view?acc_num=osu1341314153

Konaev, Margarita and Polina Beliakara, 'Can Ukraine's Military Keep Winning?', *Foreign Affairs*, 9 May 2022, https://www.foreignaffairs.com/articles/ukraine/2022-05-09/can-ukraines-military-keep-winning

Konaev, Margarita and Kirsten Brathwaite, 'Russia's urban warfare predictably struggles: fighting in cities is hard for any military', *Foreign Policy*, 4 April 2022, https://foreignpolicy.com/2022/04/04/russia-ukraine-urban-warfare-kyiv-mariupol/

Kosal, Margaret (ed), *Proliferation of Weapons and Dual Use Technologies; Diplomatic, Information, Military and Economic Approaches*, Springer Cham, 2021

Kubiv, Halyna, 'Gis Arta – So kann eine Karten-App den Krieg in der Ukraine entscheiden', *Macwelt*, 9 June 2022, https://www.macwelt.de/article/989799/gis-arta-so-kann-eine-karten-app-den-krieg-in-der-ukraine-entscheiden.html

L

Lazareva, Inna, 'WhatsApp Finds New Uses in Conflict Zones', *Reuters*, 3 August 2017, https://www.reuters.com/article/us-global-crisis-health-tech-idUSKBN1AJ0UX

Lee, Admiral His-Min and Michael Hunzeker, 'The View of Ukraine from Taiwan: Get Real About Territorial Defense', *War on the Rocks*, 15 March 2022, https://warontherocks.com/2022/03/the-view-of-ukraine-from-taiwan-get-real-about-territorial-defense

Lee, Rob, 'The Tank Is Not Obsolete, and Other Observations about the Future of Combat', *War on the Rocks*, 6 September 2022, https://warontherocks.com/2022/09/the-tank-is-not-obsolete-and-other-observations-about-the-future-of-combat

M

Maccar, David, '2 Million Rounds Headed to Ukraine from American Ammunition Manufacturers', *Free Range American*, 7 March 2022, https://www.newsweek.com/russia-spending-estimated-900-million-day-ukraine-war-1704383

Mattis, Peter, 'Contrasting China and Russia's Influence Operations', *War on the Rocks*, 16 January 2018, https://warontherocks.com/2018/01/contrasting-chinas-russias-influence-operations/

McCallum, Ken, 'Threat to UK from Hostile States Could Be as Bad as Terrorism, Says MI5 Chief', *Guardian*, 14 July 2021, https://www.theguardian.com/uk-news/2021/jul/14/public-should-be-alert-to-threat-from-china-and-russia-says-mi5-chief

McFate, Sean, *Goliath: Why the West Doesn't Win Wars, and What We Need to Do About It*, Penguin, 2019.

McFate, Sean, *The New Rules of War: How America Can Win*, HarperColins, 2019

McKinnon, Amy, 'Four key takeaways from the British report on Russian interference', *Foreign Policy*, 21 July 2020, https://foreignpolicy.com/2020/07/21/britain-report-russian-interference-brexit/

Miyata, Francis, The Grand Strategy of Carl von Clausewitz, 26 March 2021, *War Room*, US Army War College, https://warroom.armywarcollege.edu/articles/grand-strategy-clausewitz/

Monaghan, Sean and Ed Arnold, 'Indispensable: NATO's Framework Nations Concept beyond Madrid', Center for Strategic and International Studies, 27 June 2022

Morgan, Forrest and others, *Dangerous Thresholds: Managing Escalation in the Twenty First Century*, RAND, 2008, https://www.jstor.org/stable/10.7249/mg614af

Murray, W and R Millett, *Military Innovation in the Interwar Period*, Cambridge University Press, 1996

N

NATO Bulletin, 'NATO Electronic Warfare Advisory Committee Convenes in Brussels', NATO Publications, 25 November 2019, https://www.nato.int/cps/en/natolive/news_171280.htm

O

Office of High Commissioner for Human Rights, 'Plight of Civilians in Ukraine', *United Nations Press Briefing Notes*, 10 May 2022, https://www.ohchr.org/en/press-briefing-notes/2022/05/plight-civilians-ukraine

Oliphant, Roland, 'How Ukraine's drone navy is menacing Russia's superior Black Sea forces', *Telegraph*, 26 November 2022

P

Pacheco, Nicholas Paul, 'How Doctrine and Delineation Can Help Defeat Drones', *War on the Rocks*, 13 December 2022, https://warontherocks.com/2022/12/how-doctrine-and-delineation-can-help-defeat-drones

Patrick, Stewart, 'The New "New Multilateralism": Minilateral Cooperation, But at What Cost?', 18 December 2015, *Global Summitry*, Vol. 1, No. 2, pp. 115-134.

Pearson, James, and Christopher Bing, 'The cyber war between Ukraine and Russia: an overview', *Reuters*, 10 May 22, https://www.reuters.com/world/europe/factbox-the-cyber-war-between-ukraine-russia-2022-05-10/

R

Radakin, Admiral Sir Tony, *CDS Annual Speech,* Royal United Services Institute, 7 December 2021, https://www.gov.uk/government/speeches/chief-of-the-defence-staff-speech-to-the-royal-united-services-institute

Raith, Iris, 'When Arms Disappear. Europe's Persisting Problem of Disappearing Military Stockpiles', *Sphaera Magazine,* 17 October 2021, https://sphaeramag.com/when-arms-disappear-europes-persisting-problem-of-disappearing-military-stockpiles/

Retter, Lucia and others, *Persistent Challenges in UK Defence Equipment Acquisition,* RAND, 2021, https://www.rand.org/pubs/research_reports/RRA1174-1.html

Reynolds, Nick, 'Performing Information Manoeuvre Through Persistent Engagement', *RUSI Occasional Paper,* June 2020, https://static.rusi.org/20200611_reynolds_final_web.pdf

Riggs, Daniel, 'Re-thinking the Strategic Approach to Asymmetric Warfare', *Military Strategy Magazine,* Vol. 7, No. 3, Summer 2021, https://www.militarystrategymagazine.com/article/re-thinking-the-strategic-approach-to-asymmetrical-warfare

Roberts, Peter and Sidharth Kaushal, 'The Rules of Competition', RUSI, 2020, https://www.rusi.org/explore-our-research/publications/occasional-papers/competitive-advantage-and-rules-persistent-competitions

S

Shelter-Jones, Philip, 'Ten Trends for the Future of Warfare', *World Economic Forum,* 3 November 2016, https://www.academia.edu/29649157/10_trends_for_the_future_of_warfare

Shore, Jennifer, 'Don't Underestimate Ukraine's Voluntary Hackers', *Foreign Policy,* 11 April 2022, https://foreignpolicy.com/2022/04/11/russia-cyberwarfare-us-ukraine-volunteer-hackers-it-army/

Sidiropoulos, Elizabeth, 'How do Global South politics of non-alignment and solidarity explain South Africa's position on Ukraine?', Brookings, 2 August 2022

Singer, PW et al., *Wired for War: The Robotics Revolution and Conflict in the 21st Century,* Mariner Books, 2015

Small Wars, 'Liminal and Conceptual Envelopment: Warfare in the Age of Dragons', *Small Wars Journal* (interview with Dr David

Kilcullen), 26 May 2020, https://smallwarsjournal.com/jrnl/art/liminal-and-conceptual-envelopment-warfare-age-dragons

Spellar, Jhon, 'Smart procurement: an objective of the Strategic Defence Review', *RUSI Journal*, 1998

Staten, Adam, 'Russia Spending an Estimated $900 Million a Day on Ukraine War', *Newsweek*, 6 May 2022, https://www.newsweek.com/russia-spending-estimated-900-million-day-ukraine-war-1704383

Steele, Johnathon, 'Understanding Putin's Narrative About Ukraine Is the Master Key to This Crisis', *Guardian*, 23 February 2022, https://www.theguardian.com/commentisfree/2022/feb/23/putin-narrative-ukraine-master-key-crisis-nato-expansionism-frozen-conflict

Stickings, Alexandra, 'Space as an Operational Domain: What Next for NATO?', RUSI Occasional Paper, 15 October 2020, https://rusi.org/explore-our-research/publications/rusi-newsbrief/space-operational-domain-what-next-nato

Strachan, Hew and Scheipers Sibylle (eds.), *The Changing Character of War*, Oxford University Press, 2011

Strategic Studies Institute, '*Defining War for the 21st Century*', SSI annual strategy conference report, 2010

Sushko, Oleksandr 'Defending civil society in Ukraine', Open Society Foundations, 8 March 2022, https://www.opensocietyfoundations.org/voices/defending-civil-society-in-ukraine

Swed, Ori and Daniel Burland, 'The Global Expansion of PMSCs: Trends, Opportunities and Risks', OHCHR.org, undated, https://www.ohchr.org/sites/default/files/Documents/Issues/Mercenaries/WG/ImmigrationAndBorder/swed-burland-submission.pdf

T

Tegler, Eric, 'Russia may be showing it is running low on precision guided munitions', *Forbes*, 24 March 2022, https://www.forbes.com/sites/erictegler/2022/03/24/from-debuting-hypersonic-missiles-in-ukraine-to-hinting-at-chemical-weapons-russia-may-be-signaling-its-short-of-munitions/?sh=21ba7480632a

Thomas, Richard, 'Russo-Ukraine war equipment loss ten times that of Moscow's Chechen conflicts', *Army Technology*, 23 December 2022, https://www.army-technology.com/features/russo-ukraine-war-equipment-loss-ten-times-that-of-moscows-chechen-conflicts

Tirkey, Aarshi, 'Minilateralism: Weighing the Prospects for Cooperation and Governance', Observer Research Foundation, 1 September 2021

Tondo, Lorenzo and Isobel Koshiw, 'Ukraine destruction: how the Guardian documented Russia's use of illegal weapons', *Guardian*, 24 May 2020, https://www.theguardian.com/world/2022/may/24/ukraine-destruction-how-the-guardian-documented-russia-use-of-weapons

Tooze, Adam, 'The Second Coming of NATO', *New Statesman*, 18 May 2022, https://www.newstatesman.com/international-politics/geopolitics/2022/05/the-second-coming-of-nato

U

UK Cabinet Office, 'Global Britain in a Competitive Age: The Integrated Review of Security, Defence, Development and Foreign Policy', 16 March 2021, https://www.gov.uk/government/publications/global-britain-in-a-competitive-age-the-integrated-review-of-security-defence-development-and-foreign-policy

UK Govt Publications, *Armed Forces to Be More Active Around the World to Combat Threats of the Future*, 23 March 2021, Gov.UK, https://www.gov.uk/government/news/armed-forces-to-be-more-active-around-the-world-to-combat-threats-of-the-future

UK Thoughts on Defence, 'Retiring Sunset Capabilities in the Integrated Review', onukdefence.co.uk, 12 March 2021, https://onukdefence.co.uk/military-capability/retiring-sunset-capabilities-in-the-integrated-review-you-have-to-trust-someone

UK Government, 'Defence Equipment Plan, 2022 to 2032', https://assets.publishing.service.gov.uk/government/uploads/system/uploads/attachment_data/file/1120332/The_defence_equipment_plan_2022_to_2032.pdf

UK Ministry of Defence, 'Joint Expeditionary Force (JEF) – Policy Direction', Policy Paper, 12 July 2021, https://www.gov.uk/government/publications/joint-expeditionary-force-policy-direction-july-2021/joint-expeditionary-force-jef-policy-direction

V

Vergun, David, 'Near Peer Threats at Highest Point since the Cold War, DoD Officer says', *DOD News*, 10 March 2020, https://www.defense.gov/News/News-Stories/Article/Article/2107397/near-peer-threats-at-highest-point-since-cold-war-dod-official-says/

Vicic, Jelena and Rupal Mehta, 'Why Russian Cyber Dogs have mostly Failed to Bark', *War on the Rocks*, 14 March 2022, https://warontherocks. com/2022/03/why-cyber-dogs-have-mostly-failed-to-bark

W

Walker, Patrick, 'Challenges to the deployment of autonomous weapons system', Buckingham University Humanities Research Institute, August 2019, https://www.academia.edu/83823709/ Challenges_to_the_deployment_of_autonomous_weapons

Walker, Patrick, 'War without Oversight: Challenges to the Deployment of Autonomous Weapons', August 2019, https://papers.ssrn.com/sol3/ papers.cfm?abstract_id=3757516

Walker, Patrick, 'Leadership Challenges from the Deployment of Autonomous Weapons Systems; How Erosion of Human Supervision over Lethal Engagement Will Impact How Commanders Exercise Leadership', *RUSI Journal*, July 2021, https://www.tandfonline.com/ doi/full/10.1080/03071847.2021.1915702

Wavell Room Publications, 'Big Toys: British, Ground-Launched, Long-Range, Precision Strikes', 19 September 2019, https://wavellroom. com/2019/09/19/big-toys-the-requirementfor-a-british-ground-launched-long-range-precision-strike-capability/.

Wenger, Andreas and Simon Mason, 'The Growing Importance of Civilians in Armed Conflict', *CSS Analyses in Security Policy*, Vol. 3, No. 45, 2008

White, Olivia and others, 'War in Ukraine: 12 disruptions changing the world', McKinsey and Partners, 9 May 2022, https://www.mckinsey. com/business-functions/strategy-and-corporate-finance/our-insights/ war-in-ukraine-twelve-disruptions-changing-the-world

Whitman, Richard, 'UK's vision is confident, but success is a long way off', *Chatham House opinion paper*, 16 March 2021, https://www. chathamhouse.org/2021/03/uks-vision-confident-success-long-way

Wichmann, Jonathan, 'Our world is changing – but not as rapidly as people think', *World Economic Forum*, 2 August 2018, https://www. weforum.org/agenda/2018/08/change-is-not-accelerating-and-why-boring-companies-will-win/

Witt, Stephen, 'The Turkish Drone that Changed the Nature of Warfare', *New Yorker*, 9 May 2022, https://www.newyorker.com/ magazine/2022/05/16/the-turkish-drone-that-changed-the-nature-of-warfare

Z

Zabrodskyi, Mykhaylo and others, 'Preliminary Lessons in Conventional Warfighting from Russia's Invasion of Ukraine: February-July 2022', Royal United Services Institute, December 2023

Index

ABOUT THE AUTHORS

Former 5th Royal Inniskilling Dragoon Guards, Dr Paddy Walker is Managing Director of the Leon Group, a senior research Fellow in Modern War Studies at The University of Buckingham, an Associate Fellow at RUSI and previously London chair of Human Rights Watch. Paddy is a board member of the non-governmental organisation Article 36 and a regular commentator on the requirement for meaningful human control across lethal engagements.

Former Royal Navy Officer, Professor Peter Roberts is Founder and CEO of Aurelius Lab Ltd., Senior Associate Fellow and former Director at RUSI, Professor of War and Warfare at Ecole de Guerre, Paris. Peter is well known for his role as podcast host on the 'Western Way of War', 'This Means War', 'How to Train a Military' and recently 'Command and Control', required listening for staff colleges the world over.

We hope that you have enjoyed this book.
Please consider posting a review for future readers.

Thank you.

Milton Keynes UK
Ingram Content Group UK Ltd.
UKHW050641131123
432462UK00003B/4

9 781912 440498